human nature

TIMES BOOKS
Henry Holt and Company   New York

human nature

A Blueprint for Managing the Earth—by People, for People

James Trefil

Times Books
Henry Holt and Company, LLC
Publishers since 1866
115 West 18th Street
New York, New York 10011

Library of Congress Cataloging-in-Publication Data
Trefil, James S., 1938–
   Human nature : a blueprint for managing the earth—by people,
for people / James Trefil—1st ed.
      p. cm.
   ISBN 0-8050-7248-9
   1. Human ecology—Philosophy. 2. Human behavior.
3. Human beings—Effect of environment on. 4. Philosophy of nature. I. Title.
GF21.T74 2004
304.2—dc22   2003067548

First Edition 2004

Designed by Paula Russell Szafranski

Printed in the United States of America

1   3   5   7   9   10   8   6   4   2

To Wanda

Contents

Preface

I sat down to write this introduction at a strange time. During the third week of September 2003, the entire east coast of the United States was transfixed by the approach of Hurricane Isabel, which eventually made landfall in North Carolina and passed through Virginia and Pennsylvania on its way to Canada. By one of those coincidences that occur in real life, but that you would never believe if you encountered it in a novel, I was at the same time reading Erik Larsen's *Isaac's Storm*, a chronicle of the hurricane that destroyed Galveston in 1900. The contrast between the two situations was striking. In 1900, there was no warning of the coming disaster—the forecast for hurricane day was for a "brisk breeze." So sure were the meteorologists that they understood the "Law of Storms" and that Galveston was in no danger that they actually put pressure on Western Union not to transmit warnings from Cuban forecasters about the storm.

By contrast, Isabel was in the news for a week before it arrived—the Weather Channel guys were in hog heaven. Hourly satellite photos documented the storm's northward turn, its decline from a category five to category two. Airplanes flew into the gathering maelstrom to explore its structure. Reporters stationed themselves along beaches and provided breathless accounts of preparations and landfall. Officials activated evacuation plans, power companies deployed crews for cleanup, plywood went up on countless windows. This chain of events, I realized, provided a dramatic illustration of the future of the human relationship with nature that is the subject of this book.

The storm was not controlled or conquered—learning to control hurricanes is something we will only do in the distant future, if ever. But we have studied them, learned how they work, learned what we need to do to minimize their impact on us. Isabel was pretty destructive—a score of lives lost, property damage, 4.5 million people without power. But compared to Galveston, where as many as 10,000 people died and the city set up huge pyres at major intersections to burn unidentified corpses, Isabel was a stroll through the park. In a very real sense, the storm (or at least its consequences) was managed.

But great disruptions like Isabel—earthquakes, volcanic eruptions, violent storms—illustrate an important fact about modern life. For most of us most of the time, "nature" just doesn't enter our consciousness. So successful have we been at insulating ourselves from the harmful effects of the natural world that we have largely forgotten that they exist. Our lives have been so far removed from "nature" that we seldom think about what it should mean to us and how we should interact with it. And even when we do, we tend to think of the world in a way that is profoundly out of tune with the way the world—and "nature"—really are.

I suppose I began to understand this discrepancy back in the 1970s, when I bought an abandoned farm near the Blue Ridge Mountains in Virginia (I was on the faculty of the University of Virginia at the time). On a blazing hot July afternoon, I walked out over my land, choosing the spot where I would build a house and raise a family. I will never forget the feeling of—it's hard to find the right word, but I guess "inhospitality" will do—I got from the land. There was no water, no shelter, nothing but the hot sun and the dry grass. Over the next few years, I built a house with my own hands, managed a fifteen-acre woodlot in order to have firewood to heat it, maintained beehives, and generally did the back-to-the-land thing that was so popular in the 1970s. During that period I came to a kind of familiarity with nature that had been missing in my (largely urban) life up to that point. I learned that nature is not good and it's not bad—it just is.

I also learned that nature is profoundly indifferent to the fate of the species *Homo sapiens*. This came home to me one night when the temperature was plunging to ten below zero Fahrenheit and I was running around piling extra insulation on my wellhouse and doing all the other things country people do in extreme circumstances. Suddenly, I stopped and looked up at the sky, at the clear, cold light of the stars. At that moment I knew in my soul what I had learned intellectually in my studies—the universe doesn't give a damn about us or any other living thing on our planet. As far as our survival and well-being are concerned, we're pretty much on our own. The more I encountered the warm, fuzzy view of nature so often portrayed in the popular press, the more I realized that something was wrong—that the nature I was reading about wasn't at all like the nature I lived with all those years.

That insight stayed with me as my career veered from the conventional track of theoretical elementary particle physics to an attempt to become a scientific generalist (an endangered species not protected by law!). The more I learned about our planet and the living things on it, the more I realized that we are at the threshold of something unprecedented. As I will argue in this book, advances in a number of different sciences will, in a matter of a few decades, change forever the way that human beings interact with nature. The historic human discontent with what nature offers by itself, our constant struggle to improve our lives and the survival of our species, has brought us to the brink of becoming managers of our planet. Over the next decades, we will have to make hard choices about how to exercise this newfound power, and one of the purposes of this book is to get people to start thinking about the kind of world they want to live in.

I lay out the basic outline of this argument, then move on to a description of the planet we inhabit, looking at issues like climate stability, extinctions, and global warming. Some of these issues are complicated, some are embroiled in a ferocious debate that (unfortunately) mixes science and politics. I have made an attempt to present what I take to be mainstream views and to raise warning flags when there is controversy. I have also tried to label my personal views clearly, especially when they are at odds with the majority of my colleagues (as they are, to some extent, on the issue of global warming). I then outline the new fields of science that I referred to above—genomics, and complexity theory and experimental ecology—and discuss how they will affect the relationship between humans and nature in the future. Finally, in the last two chapters, I take on the subject of how to make choices in this new world that science is presenting us.

In the end, though, the accumulated history and the new sciences lead inevitably to one conclusion: if nature is defined to be that which is independent of human beings, then nature will cease to exist (if it has not done so already). From now on, the distinction between "human" and "nature" will become less and less meaningful, less and less useful in thinking about the world. "Nature" will become human.

In writing a book like this, the wise author calls on many friends and colleagues to provide a sounding board and to call attention to egregious errors. With the usual caveat that any errors that remain in this book are my responsibility alone, I would like to thank Bruce Ames, Michael Foote, Iris Knell, Douglas MacAyeal, Harold Morowitz, Michael Moseley, Jeff Newmeyer, Wanda O'Brien, Ray Pierrehumbert, as well as my children Dominique, Flora, and Tomas Waples-Trefil. I would also like to thank Kim Gareiss for encouragement in the proposal stage of the project and for useful discussions. In putting together this list, I find myself in a somewhat unusual situation, because several colleagues who provided guidance and assistance as the book was being written became so dismayed by the heretical nature of some of my conclusions that they asked not to be recognized by name. You know who you are, guys, and I thank you for your help. Paul Golob and Robin Dennis provided invaluable suggestions and editing as the manuscript progressed from first draft to final form. Finally, I would like to thank Ted Pedes and the Royal Olympic Cruise Lines for providing the weeks of comfortable isolation that allowed me to concentrate on writing the book, and the staff at Kiari's Coffee in Fairfax, Virginia, for providing a haven where I could work undisturbed, as well as for all that cappuccino.

So let me start this exploration of humans and nature by telling you about an experience I had a few years ago.

I

The First Step

1

Where Do We Fit In?

It was a beautiful day in the Black Hills, one of those days when the sky was so blue and the grass was so green that it just made your teeth ache. My wife and I had pulled into the dirt parking lot at the trailhead and were getting our hiking gear out of the trunk of our car. Suddenly, two rather agitated park workers came running up the trail. "Watch out," they said, "there's a bull buffalo coming."

And there he was, ambling slowly along the side of the hill. Not deigning to notice the wondering humans, the buffalo strolled by, grazing on the lush grass. He was big, probably near a ton in weight, and we could see the muscles rippling along his back and flanks. The deep brown of his fur contrasted with the dark tree trunks along the trail; the black of his face almost matched them. Being sure to stay behind our car, we watched as he moved along the side of the parking lot and on down the trail we were planning to hike. At that moment, with the sun shining on that magnificent beast, I experienced a feeling that is probably familiar to most modern urbanites. It was a feeling of rightness, a feeling that somehow, in this experience, I was seeing the world as it ought to be, as it would be if only humanity had not decided to pursue technology and had stayed in communion with nature.

We waited about ten minutes, then started out on our hike, following the direction the buffalo had taken. Our paths seemed to move in parallel that day, and throughout the afternoon we kept an eye on our buffalo friend, being sure to keep at least two city blocks between us. My wife decided that I had earned an Indian name—Walks With Buffalo. (My own suggestion—Runs Like Hell From Buffalo—was summarily rejected.) But as the afternoon wore on, I kept coming back to that initial reaction, that purely emotional response to being in contact with an aspect of nature that's not part of everyday experience. And as I mulled it over, I began to realize that I had stumbled onto an important dilemma that faces modern humans—the dilemma of being part of nature, yet not being part of it at the same time.

After all, here I was, driving up to a trailhead in South Dakota in a car that is a technological achievement of the first order. The power in that car's computers probably exceeds the power of the primitive computers I used to write my Ph.D. thesis more years ago than I care to remember. I was wearing hiking boots that were marvels of the engineer's art, and protecting my skin with sun block created in a major chemical factory. And to what end was I applying all this technology? To go out and spend a day far from anything engineered or constructed by human beings, to get in touch with "nature." I, along with the dozens of hikers sharing the trails with me that day, was using what science and technology had produced to escape from that very same science and technology.

I am, of course, not alone in having these sorts of contradictory feelings about the world we share. Most of us want to live comfortably, enjoying climate-controlled homes and traveling about freely in private cars. At the same time, we don't want to confront the pollution attendant on drilling for oil or burning coal. We love getting away to places like the Black Hills to camp and hike and live a simple life, but we're also very happy to get back to our urban homes, with a coffee shop around the corner and all the conveniences of civilization at our beck and call. We love hiking through old-growth forests, but a stroll down Fifth Avenue also ranks pretty high on our list of favorite activities. Tons of ink have been spilled by writers trying to convince us that one or the other of these tendencies—"civilization" or "nature"—is antithetical to the good life, or to morality, or to common sense.

But the more I thought about it, the more I began to entertain a heretical thought. What, I wondered, if *both* of these types of activities are profoundly in tune with human nature? What if we are, in fact, creatures equally at home in the quiet of a wilderness area and the hurly-burly of the city? What if both the beauty of a deserted beach and the Lake Michigan shoreline, in the shadow of Chicago's skyscrapers, are places where we belong? What if, in other words, there is no essential conflict between our need for technology and our need to seek renewal in its absence? What if our ability to create "unnatural" technologies is, in fact, the most natural thing we can do?

For there is no question, from a scientific point of view, that human beings are an integral part of the great web of life that exists on our planet. Like every other living thing, we are one result of a great experiment in molecular biology that began four billion years ago in the warm waters of the Earth's oceans, when life first appeared on our planet. We depend on the workings of the great web of life that surrounds us for all of our necessities—things like clean air, clean water, and the food we eat.

The more I thought about this question, the more I realized that there was another aspect to it. Yes, human beings are part of life on our planet, but we have also had a profound effect on the workings of the planet as well. In fact, if you think about nature as something that happens in the absence of human beings, then nature has largely disappeared from the Earth. The air that buffalo and I were breathing in the Black Hills that day, for example, was loaded with molecules produced by human activities on all of the planet's continents. The same is true of the water we drank, the weather we experienced, and the food we ate. "Nature" has become, in a very real sense, "human."

So there can be no questioning the fact that we are somehow different from other living things. There are many dimensions of this difference, but the ability to understand the world around us in abstract terms (what we call science) and the ability to use that understanding to change the environment in which we live (what we call technology) is surely one of the most important. If an extraterrestrial were to visit Earth, the first thing it would notice is that there is one species—*Homo sapiens*—that dominates the environment, changing it to meet the needs of the species. Humans just aren't like everything else.

It is this duality—this being in nature but not being in nature—that is at the root of what was bothering me out there in South Dakota. There are many ways of expressing the duality: where we come from versus where we're going, how we're the same versus how we're different, how we depend on the global ecosystem versus how we control it, and so on. But to understand what all of this means to us today, we have to step back and take a broader view of both sides of the equation, of both humanity and nature.

In a sense, the rest of this book will be a detailed look at what happens when you do that. All of us are used to thinking of the world in a certain way, of approaching problems through a comfortable and familiar process. For scientists, a familiar way of dealing with something like determining the proper place of humanity in nature is to look at history, at how things got to be the way they are. The idea, of course, is that once you know this, you have a better chance of figuring out where things are going.

It is clear that in the beginning, our ancestors weren't much different from other primates. I usually picture australopithecines like the famous "Lucy," who walked around Africa three million years ago, as being a lot like modern chimpanzees (except that our ancestors walked upright). They really were part of nature, subject to its laws, not all that different from other life-forms. In this "natural" world, their children died of diseases we no longer think about and their life expectancy measured a few paltry decades. As time passed, our species evolved into modern *Homo sapiens*, but the basics of human life changed only slightly. After all, stone axes and fire (two of the first great technological achievements of our kind) don't give you much of a barrier against a hostile world. Nevertheless, our ancestors were, in a sense, "in tune" with the natural world, interacting with it in ways that we can only imagine.

From a scientific point of view, what differentiates the lives of those ancestors from our own is easy to state—they lived out their lives in a world completely governed by the iron laws of natural selection. In a world governed by natural selection, characteristics of organisms are transmitted genetically from one generation to the next, and "unsuccessful" traits (those that do not allow an organism to reproduce and pass on its genes) are weeded out in the long run. Natural selection is a slow, inexorable process, but it's the way the Earth's biosphere has developed for almost all of the planet's history. So important is this fact that I will suggest later that whether or not a system operates according to the laws of natural selection is as good a way as any of defining the term "nature."

But then, about ten thousand years ago, the situation began to change. A succession of people, probably women in the Middle East, discovered that it was possible to grow plants and harvest their food, rather than to gather what nature produced on its own. With the development of agriculture, followed by the slow buildup of technology leading to the explosions of the scientific and industrial revolutions, we gradually removed ourselves from the natural system, based on survival of the fittest, and began to construct our own world. We learned how to grow food instead of gathering what nature offered; we learned to inoculate our children against disease and care for our sick. The more we separated ourselves from nature, the less we were willing to be content with what nature offered, the more successful we became, the greater our numbers, the richer our lives. This was what I like to consider the first step—the first separation of the human race from the "natural" scheme of things. In it, the human race stepped out of natural selection and into a world where science and technology increasingly dominated our choices and our future.

There are two ways to think about this first step. One is to note that those early farmers made a fateful decision—they decided that they would not be content to live with what nature offered freely, but would find ways to extract more from the world than the world would willingly give. A modern pharmaceutical scientist developing a new medicine and a modern engineer designing a better communication system are both following in that ancient tradition. The other way to think about the first step is to note that because of it, human beings (along with those plants and animals we have domesticated) are the only living things on this planet whose development is no longer bound by the process of natural selection. Both of these aspects of the first step have profound implications for the human future on our planet.

The importance of moving beyond natural selection can't be overemphasized. The move was profoundly unnatural, for although we still depend on the natural world for many things, as our technological abilities have increased, that dependence has become more and more attenuated.

Leaving this aspect of nature behind has had a profound effect on humanity as a whole as well as on individual human beings. Our dependence on (and our awareness of) the natural world has shrunk steadily.

There is no question that our species benefited from this change. By any material measure—the number of human beings in the world, average calorie intake, area of land settled and dominated—the human race prospered beyond all possible dreams. And yet . . . and yet . . . *something* seems to be missing. The more material goods we accumulated, the more the natural world faded into the background of our lives, the more we seemed alienated from life. I'm not suggesting that all (or even a significant fraction) of modern psychological woes are a direct result of the agricultural revolution, but it certainly led to a world where the relationship between humans and nature is less clear than it used to be.

But there is good news, because advances in a number of fields of science are developing the tools that will allow us to take a second step—a step that, in its own way, is as important as the development of agriculture. This revolutionary work will, when brought together, result in a complete recasting of the relationship between human beings and nature.

Genomics

In the nineteenth century, we learned the first great secret of life—that it is based on chemistry. In the twentieth century, we learned that the instructions for carrying out those chemical reactions were coded onto molecules of DNA in our cells. In this century, scientists are starting to learn how to manipulate those instructions—to get under the hood of living systems, as it were. Genetically modified foods, new medicines, and clones are just three examples of new things coming from this knowledge and ability. Like our hunter-gatherer ancestors who discovered that they don't have to be content with the foods provided by nature, modern scientists are developing the ability to craft living things into forms more to our liking. In the future, natural selection will be replaced by human manipulation of genomes. You may approve or disapprove of humans having this ability, but you can't deny that the ability is being developed.

Experimental Ecology

There has been, historically, a feeling that ecosystems are, somehow, too complex for humans to understand and control. Confronted with complex problems, however, scientists often return to their roots as tinkerers and craftsmen. They observe, experiment, change parameters, and just plain mess around with whatever they're studying until gradually, piece by piece, they begin to get a sense—a feeling, really—about how the object of their attention works. For the last thirty to forty years, scientists have been tending plots of prairie grass, watching controlled forest patches, and counting cacti in the desert. The result is that we are coming to understand the general rules that govern ecosystems, the rules that govern things like the biological diversity in a place, or the rules that govern the roles of various nutrients in specific ecosystems. As with genomics, we're starting to "get under the hood" of ecosystems.

Complexity Theory

Complex systems are defined to be those in which there are many agents and in which the actions of one agent can depend on the actions of all the others. A stock market, where buyers and sellers engage in a perpetual dance of action and reaction, is the classic example of a complex system, as are most large ecosystems. Complexity as a science is still only a few decades old, but it is clearly the science that we will need to understand the natural systems around us. Just as the experimental ecologists are giving us the tools to understand the basic workings of nature, the complexity theorists will supply us with the mathematical ability to predict (or at least estimate) outcomes of human interventions.

Informatics

Tying these three sciences together, and underlying all of them, are the great advances in computing, data storage, and analysis that we call the information revolution. Computers allow us to store data on the thousands of variables that can affect an ecosystem, to track the effects of things like rainfall and temperature over long periods of time, and to develop huge models that predict the future development of forests, river drainages, and farmland.

. . .

A word of caution: here and in the rest of the book, I will often talk about these developments in the present tense, but you should always be aware that they are forming and developing; they are not yet complete. Some of the pieces are already in place, but others, particularly in the areas of genetic manipulation and large-scale ecosystem management, will not be ready for a while. Given the pace of previous scientific advances, however, it's hard to imagine that it will take more than twenty years for the things I'm talking about to become reality, which means that *it's not too early to start thinking about them now.*

Scientists have been assembling knowledge over the past couple of decades in genomics, complexity, and experimental ecology, and this knowledge, taken together, leads inexorably to a new view of the human relationship to nature. Its effect will be to return human beings to nature, not as participants, as our ancestors were, but as managers. In this future, the separation between humans and nature will begin to diminish, but not necessarily in ways that are either expected or welcomed by environmental philosophers. Like it or not, ready or not, we have become the caretakers of this planet.

In fact, the best way to think of our future relationship to our planet is to think of the relationship between a gardener and a garden. No gardener wantonly destroys his or her plants, but every gardener pulls out weeds. A garden is managed to meet the needs of the gardener, and in just the same way, because of the advances described above, we are acquiring the ability to manage our planet, to shape it as we will for our own benefit. This is a message of enormous hope. The Earth is not a fragile, hopeless place, forever at the mercy of some guy with a chain saw. It is a complex, resilient system that we can learn to manage.

To summarize the scientific part of my argument, then, a look at the human past and at the state of modern science allows us to identify two enormous steps, one taken long ago, one in the process of being taken. The first step took us out of the realm of natural selection, the realm in which "nature" is to be found. The second step will make us, for better or for worse, the managers of our planet. Nature will no longer be something apart from human beings, but will be, in a very real sense, "human nature."

But the message of this newfound human ability will not be welcomed universally. For one thing, the modern scientific view of the global ecosystem has not penetrated the popular consciousness. Most people believe in what I call the tenets of pop ecology—that the planet is threatened, that until humans came along the planet enjoyed a stable climate, that we are in the middle of a massive and unprecedented extinction of species, and so on. One of the first things that has to be done before we look at our glowing future, then, is to clear away this underbrush and begin thinking about the planet as it actually is. We will find that the real situation is much more complicated than the simple tenets of pop ecology would have us believe. In some cases (climate stability, for example) the tenets are simply wrong, in others (chemical pollution and extinctions, for example) the situation is complicated, and a careful analysis of the available data is needed to evaluate the claims. In the end, we will have to come to grips with the existence of uncertainty in much of our knowledge about the planet and think about making wise choices in the face of that uncertainty.

Once we have done this, we will be confronted with one of the most important questions that will be asked in this new century: given that we have the ability to manage our planet, what will we manage it for? In our day-to-day decisions, how do we use the ability to control the global ecosystem, and to what end do we use that ability?

This is not a scientific question. Our new areas of knowledge, like all science, tell us *how* the planet works, but they say nothing about how it *ought to* work. Science may be able to tell you how to get to a goal, but it says nothing about how to choose that goal. By its very nature, science ignores the ethical (some would say the sacred) nature of what it studies. The new sciences may be able to tell me a lot about that buffalo I saw, for example, but they really can't say much about what I felt when I saw it or about the way that other people feel about nature.

In fact, the evolving view of nature in the modern industrialized world is something of an aberration in human history. For most of recorded history, human beings lived at the edge of disaster. Any natural event—a flood, a drought, a sudden influx of disease—could (and did) decimate human populations. To these people, nature was not a pleasant, warm place, but a constant danger, a constant dark force in their lives. Only when nature was tamed and brought under control—in a garden, for example—could it be enjoyed.

This attitude toward nature is clearly visible in the founding principles of the United States. Reading the sermons of seventeenth-century preachers in the Massachusetts Colony, one finds the notion of the city, built and maintained by humans, as the only proper place to be. To venture into the dark, forbidding forests was not only to place the safety of your physical body at risk, but to place your immortal soul in peril as well.

All of this began to change during the romantic movement of the nineteenth century. Most scholars see this movement as a reaction to the rationalism of the Enlightenment, but untamed and uncontrolled nature suddenly became a positive thing in intellectual life. Land was set aside for national parks in a new reverence for untouched nature: paintings of the Hudson River School in the eastern United States as well as the paintings of artists like Albert Bierstadt in the West captured this view of the world. The beginnings of the modern environmental movement with men like John Muir came from this tradition. In fact, a statement from Muir made in opposition to using the Hetch Hetchy valley as a reservoir for the city of San Francisco could easily have been written yesterday:

> Dam Hetch Hetchy! As well dam for water-tanks the people's cathedrals and churches, for no holier temple has ever been consecrated by the heart of man.

To my mind, both the romantic and environmental movements are products of privilege. Only if you know that there will be enough food no matter what happens can you have the luxury of enjoying a barren desert or a forbidding mountain range. Only if you know that your children will be warm and safe can you enjoy the awesome beauty of a thunderstorm. It is no accident that the environmental movement was born and flourishes in North America and Europe, where modern technology has allowed people to forget the precariousness of human existence. If I had to characterize the attitude of many people toward nature today, words like "reverence" and "stewardship" would come to mind. Many Americans have a strong emotional bond to what we call the "environment" and a real aversion to human incursions into the wilderness.

This means that when we talk about the future, it isn't enough to talk about new areas of knowledge. We have to add an extra dimension to our science. In this case, that extra dimension would involve the criteria we should use to apply our newfound managerial skills, and we cannot derive those criteria from the science itself. Each of us has to reach deep inside himself or herself and decide how our moral or spiritual calculus will deal with our new relationship with nature. To put it as bluntly as possible, the choices we make for the goals of our ability to manage nature are essentially moral and ethical choices, while the way we achieve those goals are decisions of technical management.

I expect that the discussion about this (nonscientific) issue of the proper goals for our management will generate the most debate on the arguments in this book, so I want to emphasize that the moral position I will be advocating represents an individual opinion, not a consensus among scientists.

When I go through the exercise of asking how the planet should be managed, I come up with a very simple rule:

The global ecosystem should be managed for the benefit, broadly conceived, of human beings.

I call this the benefit-to-humans principle.

At first glance, this may seem like an unexceptional statement—after all, it is humans who will be managing the planet, so of course it will be managed for their benefit. There are, however, well-thought-out philosophical positions that take markedly different approaches, including, for example, the view that the planet should be managed to promote things like biodiversity, the survival of endangered species, or some abstract view of "nature." I'll leave the discussion of those views for later in this book and for the moment will illustrate how the benefit-to-humans principle, combined with our new management ability, would affect the way we deal with three typical environmental issues. My three examples are: (1) the distribution of water in the Klamath River Basin in southern Oregon, where I think the conventional environmental approach gets things egregiously wrong, (2) urban air quality, where I think the conventional environmental approach has pretty much got it right, and (3) global warming and the greenhouse effect, where the data do not yet give us a clear idea of how to apply the standard of human benefit.

The Klamath River Basin

You really can't understand the western United States unless you understand water. Water is scarce in most of the high plains, and elaborate legal and cultural institutions have grown up to deal with its allocation. Without irrigation, most of the area west of the one hundredth meridian (a line that, roughly, bisects the Dakotas and Nebraska) could not support agriculture of any kind. In Montana, for example, where I have spent a great deal of time, the importance of water is illustrated by a folk saying: "Whiskey is for drinking, water is for fighting."

For many people, the most unfamiliar of the concepts that surround water in the West is that of the water right. It works like this: when a settler moved in and acquired land, he also filed a claim for water rights. This gave him the right to obtain a specific amount of water each year from a particular river or irrigation canal. In the event of a drought or water shortage, state employees begin to shut off irrigation water, starting with the most recent water right. Thus, for example, someone with a 1910 water right may continue to receive his full allotment during a drought while someone with a 1982 water right may be cut off without a drop. The water right, then, is a kind of agreement between the state and individual landowners about how the area's scarce water resources will be distributed.

All this is by way of introduction to some rather extraordinary events that took place in the Klamath Basin in the summer of 2001. The Klamath River rises in a large lake in south-central Oregon, flows south into California, and enters the Pacific Ocean near the small town of Requa in California. (The lake, called Upper Klamath Lake, is something of an oddity in itself—it's thirty-five miles long, but has an average depth of only seven feet.) Before irrigation systems started to be built by European settlers in the late nineteenth century, the entire basin was a series of marshes, shallow lakes, and seasonally flooded meadows and basins, and was home to a wide assortment of birds, fish, and other wildlife.

In 1905, the federal government started the Klamath River Project, eventually draining and irrigating over two hundred thousand acres in the basin. Farmers who moved in depended on irrigation for their water, because in spite of the marshy terrain, this region is actually high desert. Like much of the West, it depends on the melting of the winter snowpack for its year-round water supply.

In the late 1980s, a series of events happened that put the basin on a collision course with the federal government. First, the annual snowpack began to drop as the area experienced a classic western drought. By 2001, the snowpack barely reached 20 percent of normal. At the same time that the drought was worsening, the Environmental Protection Agency, in an apparently unrelated action, started to put Klamath Basin fish on the Endangered Species list. Eventually, three fish were included—coho salmon, the shortnose sucker, and the Lost River sucker. On April 6, 2001, a day the local community came to call Black Friday, the Bureau of Reclamation announced that virtually no water would be available for irrigation—in effect, that the water had to be used to save the fish. With crops already planted, many farmers could do nothing but watch them wither in the field—estimates of the economic losses from the decision run into the hundreds of millions of dollars.

The ensuing legal battles were fought over dry-as-dust issues such as whether or not a federal law could take precedence over a state's grant of water rights (it apparently can). But, at least on the official level, there was no sustained debate on the question of whether it was a good idea to give fish precedence over human beings.

Applying the benefit-to-humans principle to this situation is, to my mind, pretty easy. Against the vague value to the citizenry at large in preserving an ecosystem most of us will never see, we have an identifiable community that is suffering great loss. Telling a third-generation farm family that it can no longer work land that has been in the family for a century causes more than economic loss, it shatters the fabric of human society. Had the principle been followed, the needs of the farmers would have been weighed against the needs of the endangered species and a compromise would have been found.

As it happens, the aftermath of the Klamath Basin cutoff gives ample ground for this course of action. Criticizing the Bureau of Reclamation turns out to be (if you'll pardon the expression) like shooting fish in a barrel. In February 2001, for example, the National Academy of Sciences issued a report that said, in effect, that the decision to divert irrigation water to save fish had been based on insufficient evidence. (One argument advanced by opponents of the decision: sucker fish do well in warm, shallow water, while keeping the levels of Upper Klamath Lake high provided them with water that was cold and deep.) The charges that the decision was primarily ideological in nature seemed to be bolstered.

More importantly from the point of view of the managed planet I am proposing, it turned out that the designation of coho salmon as endangered had been based on counts of the wild population, with no attention paid to the fact that salmon can be (and are) raised in hatcheries and released into rivers. On September 10, 2001, U.S. District Judge Michael Hogan in Oregon threw out the designation of the coho salmon as endangered, calling it "arbitrary." In a rare display of common sense in a nonsensical conflict, he wondered how it could be that two genetically identical salmon swimming in the same stream, one wild and the other from a hatchery, could have different legal status. Given that government workers had been videotaped slaughtering hatchery salmon to keep them from "polluting" their genetically identical cousins in the rivers, it seems reasonable to count the total number of salmon in the environment and not quibble about where they were raised.

Some footnotes: On July 25, 2001, on orders from the secretary of the interior, the headgates of the Klamath Project were opened and irrigation water once again flowed to the embattled farmers—too late for many of them, but there it is. In the fall of 2003, the final report of the National Academy was issued, and while it did not support the claims that the original decision had been based on junk science, it did hold that the amount of water taken for irrigation had little or no effect on the survival of endangered species in the Klamath Basin. The future of the water will now be worked out by the usual political process.

In this case, then, application of the benefit-to-humans principle would clearly have dictated a different course of action from what was taken and, in the process, probably saved the federal government a lot of embarrassment.

Skopje in the Winter

Skopje is the capital of the Republic of Macedonia, one of those Balkan nations carved out of what used to be Yugoslavia. (The country is bordered on the south by Greece, on the west by Albania, on the north by Serbia and Kosovo, and on the east by Bulgaria.) A few years ago, I spent a brief but enjoyable period as a visiting professor at the University of Sts. Kiril and Methodius, which is located in downtown Skopje.

Located where the Vardar River comes out of the mountains and makes a lazy turn to the southwest before heading off to the Aegean Sea, the city occupies a spot that has been continuously inhabited by humans since at least the Upper Paleolithic period, with the first settlers arriving perhaps as long as forty thousand years ago. In its time Skopje has served as a Greek colony, a Roman market town, the capital of various short-lived Slavic kingdoms, an Ottoman administrative center, and now, since 1991, as a national capital. The city's name, so odd sounding to Western ears, is derived from the Scupi, a tribe that lived here during the Roman era. (This connection hasn't been lost on local real estate developers, who have named two posh new subdivisions "Scupi I" and "Scupi II".)

If you'll allow me to get on my soapbox for a moment, I consider the fact that I have to introduce the capital of a small European country in this way to be a major indictment of American education. Let me offer myself as exhibit A for this argument. I consider myself to be a reasonably well-educated man—after all, I'm a senior academic with degrees from universities on both sides of the Atlantic. I grew up in an ethnic community in 1950s Chicago and consider myself to be pretty well informed about the ins and outs of Eastern European culture. Yet when I rummage through my mental attic for information about Macedonia, I come up surprisingly empty. I'm aware there was someone called Alexander the Great, who wept because there were no more worlds to conquer and was played by Richard Burton in the movie. I'm aware that there was something called the Byzantine Empire that by some vague medieval process at some vague medieval time became the empire of the Ottoman Turks. I can summon up a mental audiotape of Eartha Kitt singing, "It's Istanbul, not Constantinople," but that's about it. All in all, not a ringing endorsement of our educational system!

In any case, Skopje, like Los Angeles and Mexico City, is located in a basin ringed by mountains. Unlike places like Chicago, where the winds can sweep pollutants away over the flat plains, cities like Skopje have to deal with the fact that if they allow pollutants to escape into the atmosphere, those pollutants will stay around for a while.

I had a beautiful apartment in what you could call a European-style subdivision in suburban Skopje. From the third-floor balcony, I had a magnificent vista of the mountains to the east of the town—a vista I enjoyed immensely, particularly at sunset. As the autumn deepened and winter came on, however, I noticed a strange phenomenon: over a period of days, a haze would descend over the city, until the mountains became invisible. Sometimes, standing in the old Turkish fortress in the center of town, you couldn't even see the tall downtown buildings, even though they were located only a few blocks away. When this happened, you could actually *taste* the air—it left a faint, bitter impression on your tongue, something like battery acid.

The Macedonians call this haze *magla,* which is their word for "fog" or "mist." (The word for "smog" exists in Macedonian, but apparently isn't used in this context.) The choice of this word is telling, because it implies that somehow the *magla* that descends on the city is an act of God, a natural event over which humans have no control.

Well, I beg to differ. When I traveled around the city, I noticed that there was visible exhaust coming from the tailpipes of almost every vehicle on the road. Where I live, in a suburb of Washington, D.C., cars are subjected to an annual emissions inspection, so the difference in Skopje was striking. Sitting in rush hour traffic, I could watch the pollutants drifting skyward, waiting to make their contributions to the *magla.* (Having said this, I hasten to add that in a country like Macedonia, devoting scarce resources to maintaining auto-emission standards is a luxury that cannot, as yet, be afforded.)

Seeing the effects of unconstrained air pollution in Skopje brought me back to the ongoing discussions about air quality in the United States. When people began to take the problem seriously, there were serious debates about how much regulation there should be of things like auto exhaust and smokestack emissions. These debates typically pitted environmentalists, arguing for stricter controls, against representatives of industry, arguing that such controls would harm the economy.

It is actually fairly easy to use the benefits-to-humans principle in this debate, since, unlike the Klamath Basin case, the effects on humans have always been the central issue involved. On the environmental side, the issues have to do with quality of life and the health effects of breathing polluted air. On the industry side, the argument was often cast in economic terms—that requiring catalytic converters would drive up the price of cars, for example, or that removing old cars from the streets would impose a disproportionate burden on the working poor. As always, there are good arguments on both sides, and choosing involves a balancing of the two sides.

When I do this balancing, I look at the data—modern American cities, where pollution controls are in effect, versus old Communist cities like Skopje, where they are not. To me, there is no contest. The human good brought by the controls, such as the availability of clean air, far outweighs the negatives. In this case, the conventional environmental wisdom pretty much got it right.

Global Warming, the Greenhouse Effect, and All That

The scientific and political issues surrounding the debate on global warming are complex—so complex, in fact, that I will devote an entire chapter to trying to unravel them. Nevertheless, this debate provides us with a good example of a situation in which it's not easy to see how to apply the benefits-to-humans principle.

The basic problem is this: for the last couple of centuries, the industrial revolution has been driven by the burning of fossil fuels—coal, oil, and natural gas. In effect, we have been mining carbon, combining it with oxygen, and putting carbon dioxide into the atmosphere. We know that carbon dioxide absorbs radiation that the planet sends out into space and that this absorption can increase the Earth's temperature. There is a natural greenhouse effect that has been around for billions of years—most scientists give it credit for preventing the oceans from freezing, for example. The question: will the carbon dioxide we're adding alter the Earth's existing greenhouse effect and, by doing so, alter the climate in ways that are harmful to humans and the environment?

The instruments of choice for answering this question are giant computer programs known as global circulation models (GCM). They are monuments to the ingenuity and skills of their creators and are splendid examples of how the science of complexity can be applied to real-world systems. The models incorporate literally thousands of different factors, from the amount of sea ice around Antarctica to vegetation on the borders of the Sahara, and come up with predictions of future climate. Unfortunately, there is a good deal of uncertainty about how accurate these predictions are, and this fact clouds the picture considerably. The estimate at the moment is that there will be an increase in average temperatures of between 1.5 and 6.5 degrees Celsius over the next century, with a best guess of around 2.5 degrees Celsius. (To get a rough conversion of degrees Celsius to the more familiar Fahrenheit temperature scale, just multiply the former by two—thus, a warming of 2.5 degrees Celsius would correspond to about 5 degrees Fahrenheit.)

The question, of course, is what to do with this uncertain information. To lower the atmospheric burden of carbon dioxide would require an enormous change in energy technology, with concomitant human costs in terms of economic upheaval. On the other hand, if we just "let her rip" and the upper end predictions turn out to be right, all sorts of nasty consequences, from an increase in serious storms to rising sea levels, could also extract an enormous human cost.

In a situation like this, a philosophical principle like benefit-to-humans doesn't yet do us much good, because our prediction tools simply aren't good enough to allow us to distinguish between the consequences of applying or not applying it. As a research scientist, I am, of course, comfortable with saying that we need to get a better understanding of the Earth's climate system, but that comment isn't much use in the real world, where decisions about carbon dioxide emission have to be made immediately. What we can say is that when the models get better (as they surely will), we will be able to do believable cost-benefit analyses on various policy options, and at that point we can talk about applying the principle. For the moment, however, the complexity of the Earth's climate system forces us to fall back on more general arguments (such as the no regrets policy I'll discuss in my longer look at global warming in chapter 9) and leave the application of benefit-to-humans to a future time.

As these three examples show, the way that the principle is to be applied varies from one situation to the next, and there is no blanket prescription to make decisions easy. In addition, the ability to analyze specific situations can depend on the state of scientific knowledge in areas related to it.

Having said this, however, I have to return to my original point, which is that whether we decide to accept the benefit-to-humans principle or some other, the new sciences will allow us to manage the planet in accordance with it. Nature will no longer be something independent of human activities—something that operates on its own, independent of what we do. Instead, it will be some sort of mix of things we used to call "human" and things we used to call "nature." Deciding precisely what that mix is to be will, I think, be the great human project of the twenty-first century.

2

Nature and Natural Selection

What does it mean for something to be natural? This is an important question, because if we are going to think about the relation between humans and what we call nature, we had better have a pretty clear idea of what "nature" is.

The standard sort of definition—that nature is something that happens when humans aren't around—isn't very useful in this regard, simply because by this definition there is no "nature" left on the planet. Furthermore, if that's the definition, then exhortations to live in balance with nature—exhortations we often hear—are a bit contradictory. In this chapter, I want to think about nature in a slightly more subtle but ultimately more useful way. I want to look at the way that natural systems came to be what they are—a process that went on before humans were on the scene—and then use the term "natural" to apply to systems that still operate that way today. And this, of course, gets us into the theory of evolution.

For the fact of the matter is that we have a very good picture of how the world of living things developed over the past four billion years, and this gives a very good standard of comparison with which to judge our modern world. It also gives us a scientifically based way of defining "natural." Before we do that, however, let's take a few moments to summarize what we know about the development of life on our planet.

Four and a half billion years ago, our planet was a hot, lifeless ball floating in space, with no oceans and no atmosphere. Today it teems with life. The question, then, is obvious—how did we get from there to here? This question, in turn, naturally breaks down into two parts: how did life itself arise from the inorganic stuff of the early Earth and once it arose, how did it diversify into our modern complex biosphere?

One of the great scientific discoveries of the nineteenth and twentieth centuries is that life is, at bottom, a matter of chemistry. Inside the cells of every living thing on the planet, complex molecules are locking together or breaking apart in a delicate dance that leads, eventually, to the phenomenon we call "life."

In any case, the early Earth was full of the molecules that are the basic building blocks of living systems. By a process whose details scientists still argue about, somewhere in the Earth's oceans some of those molecules assembled themselves into a system that was able to take in molecules and energy from their environment, grow, and, eventually, split into two identical systems. The first cell had been born. And although scientists still argue about the exact mechanism by which this happened, we do know something about when it happened.

When the Earth first formed, it did so in a rubble-filled disk around the sun. Sweeping through its orbit, the protoplanet swept up the debris around it. For the first 500 million years of the Earth's life (give or take a few hundred million), huge chunks of rock rained down on its surface. If a big one—say the size of Ohio—hit, it would bring in enough energy to vaporize the world's oceans and, quite literally, sterilize the planet. If life had developed before this impact, it would have been wiped out without a trace. It was only after what astronomers call the Great Bombardment ended that it was possible for life to take hold. Most astronomers place this opening of the possibility of the survival of life at about 4 billion years ago.

On the other hand, we also know that by about 3.5 billion years ago, life was flourishing on our planet. We know this because dozens of species of photosynthetic bacteria (think green pond scum) have left behind fossils in old rock formations in Australia. It appears that by this time the Earth's oceans had produced copious life. I picture this period in our history as one in which the shores of the oceans, their bays and estuaries, were covered with a green slime—our most distant ancestors.

The point of this discussion is that there was a fairly narrow window during which life could have developed—roughly between 4 and 3.5 billion years ago. Furthermore, given that the first fossil life-forms appear to be part of a fairly complex community, it is likely that life developed sooner rather than later in this interval. And this, in turn, leads to the suspicion that the development of life may be due to a fairly common set of chemical reactions, once the conditions are right. (This, incidentally, is why NASA scientists are so excited about bringing rock samples back from Mars. Since that planet had an ocean early in its history, it also may have developed life, and finding traces of that life could tell us a great deal about how common the phenomenon might be in the universe.)

Whatever those first cells on Earth might have looked like, they were undoubtedly very primitive by modern standards. They probably involved little more than an ordinary set of chemical reactions, without the complex DNA-based system used by modern cells. How we got from those primitive cells to the present is another story.

Think about those early bacteria that occupied our oceans. Soon after the first group formed in the warm, energy-rich waters, the process of competition would have begun. By chance, one primitive cell would acquire something—a different mix of molecules, perhaps, or slightly different molecules in its outer membrane—that allowed it to extract energy from its environment a little more efficiently than its neighbors. Over time, this individual and its offspring would reproduce more rapidly and would quickly come to dominate the population. It's not that the old versions of the cell would be killed off. They would just become relatively fewer in number and, eventually, vanish.

At the same time, a process of dispersal would have begun. Today, storms routinely blow soil, bacteria, and even, occasionally, birds from one continent to another. Proprietors of resort communities in the Caribbean, for example, dread the storms that bring dust and sand from the Sahara into their skies, preventing their clients from getting expensive suntans. On the early Earth, the same storms would undoubtedly have spread living things around the globe, probably in a matter of a few decades. Thus, once it developed, life quickly would have become global.

The characteristics that would make a bacterium successful in the tropics, however, would not necessarily have made it successful in arctic regions, and vice versa. Different populations would come to dominate in different regions. In modern language, we would say that the original species had developed into two distinct species inhabiting two different regions. Consequently, we can say that neither the existence of life nor its diversity require any extraordinary processes—no divine intervention—to happen. It was just a matter of normal, everyday processes at work.

The Evolution of Nature

In 1831, a young English naturalist by the name of Charles Darwin began a round-the-world trip on a ship called the *Beagle*. Like the legendary *Enterprise* in the *Star Trek* series, it was on a five-year voyage of exploration. By the time the ship returned to England, Darwin had already worked out the main ideas of what was to become the main intellectual thread that ties all of the life sciences together—the theory of evolution.

A word about terms: scientists tend to be a pretty conservative lot as far as terminology goes. In everyday language, the word "theory" connotes a kind of hesitancy, perhaps a doubt about authenticity. In the sciences, however, calling something a "theory" or "law" or "principle" has nothing to do with how much evidence there is to back it up, but is a historical relic from the days when the idea was first launched. It is not at all unusual, for example, to have a "law" incorporated as one part of a much deeper and more complex body of thought that carries the name "theory." Newton's *Law* of Universal Gravitation, for example, is one part of Einstein's *Theory* of General Relativity. The fact that evolution is called a "theory," then, shouldn't blind you to the fact that it is one of the best thought out and most thoroughly verified bodies of thought in all the sciences.

Here's another way to think about this issue: there are many theories of gravity—that of Newton, that of Einstein, and, someday, what scientists will call a "Theory of Everything." There is also, however, the fact of gravity. Drop something and it falls. Period. In the same way, there may be many variations on the theory of evolution, but there is also the fact of evolution, just as there is the fact of gravity.

Darwin begins his great book *Origin of Species* by pointing out an obvious fact. Human beings have, over the past few millennia, produced an astonishing variety of domesticated plants and animals. Consider, for example, dogs like the Chihuahua and the Saint Bernard. Both descended from wolves that were domesticated by our hunter-gatherer forebears many thousands of years ago and both acquired their present form as a result of selective breeding. If you want a small, hairless dog, you breed the smallest and least hairy dogs you can find, then select the smallest and least hairy from the resulting offspring to breed again, and so on. Keep this up for a while and you wind up with a Chihuahua. Apply different standards of selection and you wind up with the Saint Bernard or Great Dane. All of the foods we eat—plants and animals alike—are the result of thousands of years of selective breeding. Darwin called this process artificial selection, and one of the most charming parts of his book is his long discussion of the different kinds of domestic pigeons that it has produced (he was a pigeon fancier and apparently socialized with local men with the same interest).

After going at length into the various applications of artificial selection, Darwin asked one of those questions that, in retrospect, seem simple, but in fact are the hallmark of true genius. If humans can produce biological variety by conscious selection, can nature do the same without being conscious of it? It was this question that launched the life sciences on a trajectory that took them away from a dry cataloguing of different varieties of living things to a branch of science with true predictive and explanatory power.

The key notion in Darwin's theory is "natural selection," a term he coined to make an analogy with artificial selection. Instead of conscious human choice as the main operating mechanism for natural selection, however, he substituted two statements:

1) There will always be differences between individuals in a population.
2) There will always be more individuals in any population than can be supported by the environment.

These two statements—so obvious to us now—are the basis on which life developed from that first green pond scum to the enormous diversity we see around us today. Let's look at a simple example to see how they work. Suppose we have a population of rabbits. Just by chance, some rabbits will be able to run a little faster than others. Today, we understand that such differences have to do with the genes that the rabbits inherit from their parents. Darwin, of course, knew nothing of genes or DNA, but he knew that there was some mechanism by which characteristics are passed from one generation to the next.

In any case, if the rabbits live in an environment that includes predators—foxes, for example—the ability to run fast has a certain utility. The faster rabbit has a better chance of evading the fox and therefore of living long enough to have offspring. In the same way, the slower rabbit is more likely to be caught. Thus, the next generation of rabbits will be produced in disproportionate numbers by rabbits who can run fast. Put another way, the "fast" genes are more likely to make it into the next generation than the "slow" genes. Over time, from one generation to the next, this same scenario will play out, and, gradually, the population of slower rabbits will disappear. In the language of evolutionary theory, we say that the genes that produce fast rabbits are "selected," and that the faster rabbits are more "fit." The phrase "survival of the fittest" is often used to describe this process.

There are several points that can be made about this type of natural selection. First, although we use the word "select," in fact there is no conscious actor choosing certain traits, as there was in the breeding of the Saint Bernard. The selection is made by the environment. The second point is that it is not necessary that every fast rabbit survive and every slow rabbit get eaten. In the real world, bad luck or injury could easily consign a fast rabbit (along with his or her genes) to the fox's digestive tract. It's quite possible, for example, that the roll of the dice might actually produce a slower generation once in a while. It's just that *on average* the fast rabbits have a slight edge, so over long periods of time, you expect the selection to take place. Think of a roulette game as an example. Occasionally, people can, and do, win lots of money by "beating the odds." Over time, however, the house always wins—that's why casinos can stay in business. In the same way, over time, the fast rabbits will do better.

It is important to note that fitness is completely defined by the environment, and what constitutes fitness in one setting might be irrelevant in another. For example, in an environment without predators, there would be no selection for speed, and the prize might go (for example) to rabbits that could breed faster.

Because the concept of "survival of the fittest" acquired a bad name when some Victorian writers tried to use it to justify class distinctions in England, it might be a good idea to explore it in a little more detail. As you can see from the above example, "fitness" is defined by one thing—the ability to move genes into the next generation. Thus, "fitness" has nothing to do with how much economic success an individual has, but only with whether that individual is successful in producing children. I often use the example of Leland Stanford to make this point. Stanford was one of the great industrialists (some would say robber barons) of the late nineteenth century. Working his way up from a grocery store in post–Gold Rush California, he built the Southern Pacific Railroad, drove the spike at Promontory Point, Utah, that completed the first transcontinental rail line, and amassed a fortune sufficient to found the university of which I am proud to be a graduate. Yet he had only one son, who died in late adolescence and in whose memory the university was founded. (The official name of the school is actually Leland Stanford *Junior* University.) In Darwinian terms, Leland Stanford, for all his wealth, was unfit. His genes disappeared from the gene pool, and that's all that natural selection cares about.

Finally, you have to understand something very important about natural selection: it has no direction and no purpose. This, in the end, is what I believe bothers many people about the theory. There is no purpose, no direction, no ultimate good to the way life developed. If the environment favors fast rabbits, fast rabbits will eventually appear. If the environment changes, other kinds of rabbits will show up. There is no invisible hand behind evolutionary change, inexorably producing intelligence or moral goodness. Evolution just is.

Because I will use the ideas behind natural selection to define the term "natural" later in this chapter, I would like to take a few moments to talk about the evidence for the theory. Some of this will be (I hope) a review for many readers, but I think that it's a good thing, every once in a while, to think about how it is that we know what we know. Basically, there are three lines of proof for the theory of evolution: (1) the fossil record, (2) the evidence of DNA, and (3) imperfections in living things.

The Fossil Record

When a plant or animal dies, it can happen that it gets buried and separated from the biosphere. It may sink to the bottom of the ocean and be covered with sediment, for example, or be buried in a flood. In this case, the usual processes that serve to scatter the organism's remains won't operate, and an infinitely more interesting fate may be in store for it. Over time, as sediment accumulates above it, minerals in the water flowing through its burial grounds will be exchanged, atom by atom, with the calcium or other atoms in the organism's hard parts. The eventual product, after hundreds of thousands of years, will be a fossil—a reproduction in stone of the hard parts of the original living thing. It is in this way that the Earth remembers the living things that once roamed its surface.

The term "fossil record" refers to all of the fossils that have been found and catalogued by scientists over the last few centuries. The most dramatic fossils, of course, are those gigantic skeletons of dinosaurs we see in museums, but the real bread and butter of the fossil world are much more mundane things—buried clam shells from the ocean margins, imprints of leaves in mud flats, and, of course, the imprints of the green pond scum that are the earliest known life-forms on our planet. It is in these humble organisms, rather than the grand giants, that scientists read the story of the development of life.

Starting with simple single-celled organisms, the record shows a general increase in the diversity of living things over time. About a billion years ago, more complex cells developed, with DNA packed into a nucleus: about eight hundred million years ago, multicelled organisms appeared; four hundred million years ago, life moved to land; between eight and four million years ago, the first "humans" appeared on the scene. The fossil record allows us to trace out the intricate family tree of living things over this vast sweep of time. Name an organism and the record can tell you (in principle, at least) where it came from and how it got to be what it is.

I don't, however, want to give you the impression that the story of life on our planet is an uninterrupted triumphal march to the present. In fact, the story of life is punctuated by a number of events known as mass extinctions, in which anywhere from 30 percent to 90 percent of all species on the planet simply disappeared. The extinction of the dinosaurs, along with about two-thirds of all the other species on the planet at the time, is the best known of what paleontologists call a mass extinction event. There are many of these events in the record, and the one involving the dinosaurs is neither the most deadly nor the most recent. Mass extinctions seem to occur about every twenty-six million years and seem to act as a kind of massive reshuffling of the deck for life-forms. When the dinosaurs disappeared about sixty-five million years ago, for example, the way was cleared for mammals to take over the planet.

In terms of natural selection, both the periods of normal change between mass extinctions and the response to the extinctions themselves make perfect sense. I think of this as the orchid versus the cockroach problem. During normal times, when climate and the environment are changing relatively slowly, organisms that get very efficient at exploiting a very narrow ecological niche can flourish. These organisms are following what I call the orchid strategy, using the fact that relatively stable ecosystems reward specialization. During events like asteroid impacts, however, the organisms that do best are those like the cockroach, which are flexible and good at dealing with unfamiliar situations. Times of rapid change, in other words, reward generalists.

In the end, then, the fossil record gives us a kind of family tree, a story about how life developed on our planet. Like any family tree, it tells us how closely things are related. Two organisms on branches that split only recently, like humans and chimpanzees, are closely related, while those on branches that split a long time ago, like humans and rattlesnakes, are less closely related.

The Story of DNA

Another, more modern way of categorizing living things is to look at their DNA. The DNA of every organism carries the blueprint for running that organism's chemical reactions. If you think of the familiar double helix of the DNA molecule as a sort of twisted ladder, then the DNA code is contained in the molecules that make up the "rungs." By a relatively straightforward but somewhat complicated process, this code is turned into the chemical processes that make each organism unique. It is the DNA code, then, that makes one organism different from another within a given species and that also makes one species different from another. (I should add that it is the newfound ability to read and manipulate this code that forms one of the bases for my claim that humans will soon have the ability to manage our planet.)

During the past decade, we have learned how to read the DNA code of living organisms. The great milestone in this story occurred in 2000, when scientists announced the first assembly of the entire human genome, but the first entire genome of a free living organism (the bacterium *Haemophiles influenzae*) was done a few years earlier, in 1995. Today, the process has become so automated that people scarcely notice when the readout of the DNA of a new species is completed. As I am writing this, for example, the scientific journals are announcing that the genome of the *Anopheles* mosquito has been completed. This is an important development, because this is the mosquito that spreads malaria. Despite this importance, however, the story was not carried with any prominence in any of the major news media—I know I didn't see anything about it. The whole genome decoding process has become mundane.

But the availability of information on genomes gives us another way of constructing the family tree of living things. Consider two organisms that have a common ancestor. The longer it has been since that common ancestor lived, the more time there has been for changes in DNA to accumulate. Organisms whose DNA carry similar codes, then, shared a common ancestor fairly recently, while those with widely divergent codes have to go further back to get to a common ancestor.

Actually, we haven't yet accumulated enough data to do a full family tree of every living thing, although I expect that eventually we will do so. Nevertheless, we do have enough bits and pieces of information to see roughly how the tree works. In cases of special interest to scientists, detailed analysis of parts of an organism's DNA have been done to untangle complex genealogies. It was this sort of detailed analysis, for example, that showed that humans are more closely related to chimpanzees than to gorillas, with the gorillas branching off first, followed by a split between humans and chimps between eight and twelve million years ago.

But the key point is this: when the family tree is constructed from the available evidence of the DNA, *it is the same family tree that comes from the fossil record!*

I want you to think about this statement, because it's really rather extraordinary. If the theory of evolution is true, then this is, of course, exactly the result you would expect. It is possible, however, that things could have turned out differently. Human DNA might have turned out to be closer to that of a frog than that of a chimp. Had this happened, we would have had to discard the theory of evolution. The fact that it didn't happen, then, not only supports the theory but, in the language of philosophers, shows that the theory is falsifiable. One of the characteristics of a real scientific theory is that you can imagine results that would prove it wrong, whether those results ever actually show up or not. One of the major faults of theories like creationism and its modern incarnation, intelligent design, is that they are not falsifiable. No possible evidence could ever come to light that could prove them wrong, because the response to any negative result is just "Well, that's the way things were made."

Imperfect Design

Why is grass green? I don't want the standard schoolbook answer— "because it contains chlorophyll." I want you to think about grass in a deeper way. After all, a blade of grass or a leaf is, essentially, a solar collector, whose purpose is to gather the energy in sunlight and convert it into the carbohydrates that are the basis of the food chain on our planet. So one way to think about my question is this: what color *should* a solar collector be?

Obviously, the most efficient collector would absorb all the energy that falls on it, reflecting nothing. It would be black. Any intelligent engineer would have made grass black, and any student who was told to design a solar collector and came back with something green would surely get an F. So how did natural selection provide us with a green world?

The basic point is that natural selection can only work on what is available at any given time. Thus, if for some (as yet unknown) reason the chemistry involved with chlorophyll gave some of that early pond scum an advantage over its neighbors, then it would have been selected, and future evolution would only involve choices between organisms that shared that basic chemistry. And that unknown reason need not have been deep—it could be something as simple as the fact that chlorophyll-based chemistry appeared on the scene first. In the end, though, evolution doesn't produce perfection; it just produces survival.

There is actually a familiar everyday example of this process, the so-called QWERTY keyboard, named after the letters in the top row of any keyboard. In the nineteenth century, the first typewriters were cumbersome affairs, and the keys actually hit the paper underneath the machine. If two keys got jammed together, freeing them was a laborious process. The modern keyboard was designed to slow down the typing to keep keys from getting jammed. In the language of evolutionary theory, in the environment of the nineteenth century, the QWERTY keyboard enjoyed a selective advantage and beat out its competitors. From that point on, virtually all keyboards used this design.

So why do we still use it today, when we use the keyboard to move electrons instead of metal keys, and there's no danger of jamming anything? One way to answer that question is to ask yourself whether you would be willing to invest the time and energy into learning a new keyboard; another is to think about the cost and inconvenience of converting all of the billions of keyboards in the world to a better design. For better or worse, QWERTY has been incorporated into our technology and won't be dislodged. In just the same way, green grass, even though it's inefficient, is here to stay.

I often tell my students a story to illustrate this aspect of evolution. Two hikers are walking in the woods when they encounter a very angry (and very hungry) grizzly bear. One hiker starts shedding his knapsack and other paraphernalia.

"What are you doing?" his partner asks.

"I'm going to run," is the reply.

"Don't be silly. You can't run faster than that bear."

"I don't have to run faster than the bear—I just have to run faster than you."

Every living thing on Earth is the progeny of organisms that were "good enough"—that could run faster than the other hiker—rather than of organisms that were perfectly designed. There are many examples of these imperfections. Think, for example, of the human appendix, the vestigial leg bones buried in the blubber of whales, and the panda's thumb as examples. (If you haven't read the late Stephen Jay Gould's essay "The Panda's Thumb," contained in a book of the same name, you have a real treat in store for you.) Paradoxically, examples of imperfections in the design of living things like these provide some of the most dramatic underpinnings for evolution.

With this background about the theory of evolution, and with some confidence that the theory is firmly rooted in facts about the world, we can return to the questions with which we began this chapter. What is "nature"? What is "natural"? The hallmark of the way nature operated before humans came on the scene is the process of evolution through natural selection. I would like, therefore, to propose the following definition of "natural":

A system is natural if the living things in it develop according to the rules of evolution by natural selection, whether or not the environment in which those living things exist is altered by human beings.

The point of this definition is that it allows us to speak of natural systems in a world dominated by human technology, and it also is in keeping with the way the word "natural" is generally used. Natural selection, after all, tells us how living things respond to their environment, not how the environment came to be the way it is. We can, for example, speak of the ecosystem of Yellowstone National Park as being natural, even though there are humans there and the products of industrial civilization come into the park in the air and water. We can also speak of the development of antibiotic resistance by bacteria as a natural process, even though human beings altered the bacterial environment by introducing medicines.

An Iowa cornfield, on the other hand, would not be natural, since the reproductive success of the plants depends on whether or not human beings decide to breed them, not on their ability to survive on their own. This decision is made on a complicated economic basis—does this particular type of corn produce profits for the farmer? Do its requirements fit with the operation of the farm? Is it more profitable than other brands? None of this has anything to do with natural selection, and this definition of nature allows us to brand the field as unnatural—another example of the definition being in tune with general usage.

In addition, this definition of "nature" also leads to an interesting question: to what extent is human society natural, and, insofar as it is not, what is the process by which we pulled ourselves out of the natural system? It is to this question that we now turn our attention.

3

Leaving Nature Behind

Beings that we could call "human" have walked the surface of our planet for about four million years, give or take a couple of hundred thousand. Our most distant ancestors surely lived in a world that was natural, a world dominated by the iron laws of natural selection. Those with poor immune systems died early and their genes gradually disappeared from the pool. The ancestors of modern leopards and eagles culled the weak, the lame, and the sick. Then, if ever, life was nasty, brutish, and short.

We know about these distant ancestors from the fossil record. Perhaps the most famous of these fossils was discovered in Ethiopia in 1974. Called Lucy (her discoverers celebrated their find by sitting around a campfire listening to the old Beatles song "Lucy in the Sky with Diamonds"), she and her kind lived on the African plains about three and a half million years ago. They were an example of what scientists call australopithecines ("southern apes"). Lucy was a remarkable find because we have most of her skeleton from the neck down—what paleontologists call the "post cranials." She stood perhaps four feet tall and was probably covered with hair, like a modern chimpanzee. She had a brain with a volume of about 400 cubic centimeters—about the same as a chimpanzee or a newborn human.

(For reference, the average brain size of a modern *Homo sapiens* is about 1400 cubic centimeters.) Lucy and her kin lived in groups and, most importantly, they walked upright. Their hands, in other words, had been freed to carry out tasks other than locomotion. But at the end of the day, they were just another kind of primate, living in social groups and relying on their wits to evade the dangers in their world.

So how did we go from this rather inauspicious beginning to a species that dominates the global ecosystem? Perhaps the best way to visualize the answer is to imagine a sidewalk about two hundred yards long—the length of a long city block—with one end starting at Lucy and the other representing modern times. (Although the oldest known hominid fossils date back seven million years, Lucy is a convenient starting point for beings we could call human.) Let the height of the sidewalk above ground level represent human technical prowess—human ability to alter and control the environment.

If we start at the early end of the sidewalk, we will walk over basically level ground for almost half the length of the block. During this early stage, many different kinds of *Australopithecus* came and went on the plains of Africa, leaving behind little more than a sparse fossil record that later humans would find and interpret. Just before the halfway point—about two and a half million years ago—the sidewalk makes a slight uptick as a new species with a new ability appears on the scene. The fossils of *Homo habilis* ("man the toolmaker") are found associated with the first stone tool—the hand ax. A hand ax is just a fist-sized stone that has been chipped along one edge so that it can be used as a cutting tool. It would be the staple tool of our ancestors for over a million years. *Homo habilis* had a brain size of about 750 cubic centimeters, a significant increase over little Lucy.

A few steps farther on the sidewalk makes another uptick as a closer ancestor—*Homo erectus* ("man the erect")—appears on the scene. Standing as tall as modern humans, endowed with a brain size ranging from 900 to 1100 cubic centimeters, he makes more complex stone tools, moves out from Africa to Asia and the Pacific, and, most importantly, domesticates fire. From here on the sidewalk continues to slope incrementally upward, yet it is clear that, as important as these advances were, *habilis* and *erectus* still lived in natural systems. Their lives and their fortunes were still dominated by natural selection, just as Lucy's had been.

Walk farther. Some ten yards from the end of the block—a couple of hundred thousand years ago—the first of what paleontologists call anatomically modern humans begin to appear on the scene. You may know them as Cro-Magnon man, and, dressed in modern clothes, they would attract no attention in a New York subway car. They were us. They developed language, more complex tools, and musical instruments. The sidewalk is sloping up a little more now.

But when you get to within a few feet of the end of the block—a time corresponding to about ten thousand years ago—the sidewalk suddenly angles sharply upward. You have to strain to take this last step. This point marks the beginning of what I have called the first step, the time when humans began to cultivate crops and raise animals rather than forage and hunt for them.

A few inches from the end of the block, the sidewalk suddenly rears up precipitously, becoming almost vertical. This is the culmination of the first step, the beginning of modern science, the industrial revolution, electrification, information technology, and everything else that makes modern society what it is.

Climb that vertical wall, sit on top of it, and look back over the long struggle of our kind to achieve our present state. For most of those long millennia, our ancestors lived in the natural world, a world dominated by natural selection. Somewhere in that last couple of feet, however, we took ourselves out of that world—out of the process of natural selection—and became something unique in the history of our planet. This chapter is an exploration of how that rather astonishing thing came to pass.

The Agricultural Revolution

I don't like the word "revolution," at least as it applies to science and technology. It implies a kind of clean break with past traditions and achievements and almost never accurately reflects real events. The cartoon version of the dominating shift that I have called the first step, the agricultural revolution, for example, has a woman somewhere in the Middle East walking through a field of wild grains and having an "Aha" moment. She realizes that she can plant and tend the grain, and that if she does, her tribe will no longer have to wander around searching for food. She explains this to the tribe, which immediately settles down to raise crops and—presto—agriculture is born.

Of course, a little thought will convince you that it couldn't have happened that way. For one thing, agriculture is not a single technology, a single good idea. It is, in fact, a whole constellation of technologies. A farmer, for example, needs to be able to (1) find seeds, (2) prepare the ground for planting, (3) cultivate the crops, (4) protect the crops from various kinds of predators, (5) harvest the crop, and (6), perhaps most difficult in a society without electricity and fungicides, store the food until the next harvest. The chance of all of these types of knowledge developing simultaneously is pretty small. Just as the beginnings of the modern transportation system involved a complex interweaving of the development of cars, the building of highways, the refining of petroleum, and many other things, so, too, did the beginning of agriculture have a more nuanced story than the simple cartoon.

In fact, we can approach the question of the origin of agriculture by asking three questions:

1) *Where* did it happen?
2) *How* did it happen?
3) *Why* did it happen?

The "where" part of the story is pretty straightforward. On an arc stretching from Israel to Iran, hugging the flanks of the Taurus Mountains, is a region archaeologists call the Fertile Crescent. It is here that they have found stone sickles (whose wear patterns indicate a regular harvesting of wild grains), the remains of wheat with larger kernels than the wild varieties (which indicates the beginning of plant breeding), and other stone tools obviously suited to agricultural pursuits. It is also in this region that they have found the oldest permanent human settlements—the stone villages of Jericho on the Jordan River and Catal Huyuk (pronounced Sha-TOHL hoo-YOOK) in Turkey. That agriculture should have taken hold and flourished in this area is not surprising. As the glaciers of the last Ice Age receded, this region was warm and rainy. It was also the natural range of a number of the wild grains, mainly wheat and barley, that formed the basis for farming.

Another fact—one often ignored in the discussion of the origins of agriculture—involves simple geography. The Taurus mountain range is one of the few mountain ranges in the world that combine a temperate climate and a piedmont (what archaeologists call hilly flanks) that runs east and west.

To see why this is important, imagine a group of early humans trying to develop agriculture on the east coast of North America. If they were in a climate zone that favored agriculture, their population would grow and their settlements would expand. To move either north or south would take them into different climate zones. To flourish there, they would have to develop new technologies—cultivate new plants, devise new strategies for housing, and so forth. If they expanded to the west, however, they would stay in the same climate zone and could simply start anew using the technology they already had.

In North America, however, the westward movement would quickly run into mountain ranges, and moving to higher altitudes would involve the same sorts of difficulties as moving north or south. In the Fertile Crescent, however, farmers could simply move across the face of the mountains, which extended for hundreds of miles in the same climate zone. It was a situation tailor-made for expansion.

How we made the leap to agriculture is a little more difficult to explore, because humans who make their living by foraging (i.e., by hunting and gathering wild produce) leave little behind for future archaeologists to study. By observing modern foraging societies and thinking about the way wild food can be gathered, however, we can at least put together a plausible story about how the transition occurred.

As I mentioned in the preface, during the 1970s, I bought an abandoned farm in the Blue Ridge Mountains of Virginia. On this land I built a house with my own hands, heated it with wood from my own land, kept bees, and, in general, did the whole "back to the land" thing. I quickly became familiar with the plants on my property and began to modify the environment to encourage the ones I considered most useful. There were, for example, several old apple trees that had been planted decades before by previous owners. I cleared the brush and honeysuckle away from these trees, trimmed away their deadwood, cut down neighboring pines that were shading them, and fertilized their roots. My reward: copious yields of apples every summer and fall.

There was also a stand of elderberry bushes, and I quickly learned that these plants were extremely valuable. Not only could you make two different kinds of wine from them (one from the flowers, one from the berries), but they made delicious jams and jellies. Even the hollow stalks were useful—I used them as taps in an (ultimately unsuccessful) attempt to get syrup from my maple trees. Because of this usefulness, I once again set out to encourage the plants. I cleared brush from around them and transplanted seedlings to other likely spots on the property. Again, I was rewarded by seeing the plants (and my wine cellar) flourish.

Imagine my surprise upon starting to research this chapter, then, to learn that these commonsense activities are what archaeologists call protofarming or complex foraging. They amounted to an attempt to modify the environment, to release some plants from the bonds of natural selection because I perceived them to be useful. In the same way, archaeologists argue, the agricultural revolution was preceded by a long period during which human groups mixed a hunting-gathering strategy with the kind of environmental management outlined above.

If you think about the advent of agriculture in terms of protofarming, it's not hard to imagine a long period during which people who were essentially hunter-gatherers slowly developed the skills needed for a more settled existence. A particular stand of grain or a particular grove of trees might first be protected against competitors, for example. Children might be sent out to keep birds away from ripening fruit. Seeds from a particularly good stand of wild wheat might be carried and planted somewhere in good soil. Over time, more and more of this kind of activity could be interspersed with older hunting-gathering behaviors. Eventually, some groups decided to use one place (one with a particularly good water source, for example) as a base from which to forage. Some scholars have suggested that this run-up period to agriculture might have lasted tens of thousands of years before the "revolution."

Of course, it's not easy to find evidence for this sort of behavior in the archaeological record. Nevertheless, some traces of a preagricultural period of complex foraging behavior (including some primitive settlements) have turned up recently. The story of wheat is particularly interesting. In the wild grain, the competitive advantage will go to plants with small kernels that ripen quickly and are loosely attached to the stalk so that they can be blown off and carried away by the lightest breeze. For humans, however, the valuable plants are those with large kernels that will stay on the stalk through the rough and tumble of harvesting. Archaeologists have found evidence for larger kernels in the debris of ancient preagricultural settlements, indicating a process of (conscious or unconscious) human intervention in the environment.

The settled agricultural life seems so normal to us that it's hard to realize that there is a serious question to be answered about the agricultural revolution. The question, basically, is why anyone in his or her right mind would give up the (relatively) easy life of the hunter-gatherer to take up the backbreaking tasks of the farmer. Even today, the life of hunter-gatherers seems pretty relaxed—it's rare for people in these cultures to put in more than a few hours of work a day, and there is ample time for social interactions. In addition, the records we have indicate that after the advent of agriculture, humans declined in stature and exhibited a higher incidence of disease than their nonagricultural predecessors. Not only was farming no fun, it wasn't even particularly healthy.

Traditional theories about the advent of agriculture fall into two classes—those that ascribe the move to changes in the environment, and those that ascribe it to changes in human culture. (And, of course, academics being what they are, there are a number of theories that combine these two.)

For my purposes, which (if either) of these two extreme schools of thought is right doesn't matter very much. Despite the apparent disadvantages associated with the settled life, there is a simple argument that shows that people who start agriculture acquire a Darwinian advantage over their hunting-gathering neighbors. The basic point is this: existence for our ancestors was a constant battle between our need for calories and nature's unwillingness to give them up. It is well known that in predator-prey relationships, the population of predators fluctuates wildly with the availability of prey. Let a disease decimate the rabbits in a given area, for example, and the population of the lynxes that feed on them will fall. Furthermore, if that population falls below a certain level, it won't recover, and the lynxes will disappear from the region along with the rabbits.

It's a little like a balky car engine starting on a cold morning. Its activity fluctuates up and down, alternately racing and skipping. If there is too steep a downward fluctuation, however, the engine dies. In the same way, if a population of prey drops too low, the predators disappear.

People who depend on hunting and gathering are completely dependent on what nature has to offer. A few years of drought, or the appearance of insect pests that destroy a crucial food resource, and they're in trouble. The best defense against the sort of fluctuation into extinction described above is a large population. A hunting-gathering society, however, cannot increase its numbers arbitrarily. Nature will only provide so much, and an attempt to increase the human population will necessarily run up against that limit. History is full of examples of early humans hunting large game animals to extinction (think of the mastodon). A hunting-gathering society, then, must of necessity have a low population density and remain vulnerable to downward fluctuations. Wait long enough, and it is a mathematical certainty that there will be an extinction in any given area, an extinction that can't be avoided by moving to another territory and displacing groups already living there.

Agriculture—even the kind of protofarming described above—increases the resource base available to a population and therefore allows the population to increase, giving it a hedge against the threat of extinction. Unlike hunting, where increased exploitation of the resource tends to decrease the population of prey, thereby putting the hunters at risk, farming can be expanded pretty much at will. Population and cultivated land area, then, will increase hand in hand. Over time, farmers have a better chance of survival, no matter how hard their life is. Farming, in other words, offers a better survival strategy than hunting and gathering. In the end, farmers win the Darwinian game.

Of course, once this process starts, there is no turning back. The high population density of the farming community quickly passes the level that can be supported by the surrounding habitat, and from that point on, the agricultural strategy is locked in. Humans have no choice but to work the land, to give up the easy way of the forager. Ten thousand years ago, by whatever process, for whatever reason, our ancestors took an irreversible step. They started to demand more from nature than nature would willingly supply, and they began to modify their environment to meet their demands.

All of the technology developed up to that time—all of the different stone tools, even the domestication of fire—really amounted to little more than the human equivalent of a beaver dam or an anthill. With agriculture, however, humanity moved past the simple modifications of the environment seen in other species and embarked on a grand experiment. For the first time, a species dared to step outside of the evolutionary game, dared to demand that the iron law of natural selection be repealed. Alone among the living things on our planet, our ancestors began to build a new kind of life for themselves.

In the beginning, of course, the changes were small and incremental. Disease was still rampant, and strong immune systems still counted. But as people moved into settled communities, becoming seriously ill was no longer a death sentence. You could be cared for, and your temporary immobility no longer posed a threat to the entire group. You might survive and pass your genes to the next generation. Human social structure and technological prowess began to trump natural selection.

The Industrial Revolution and Beyond

The process that began in the Middle East in the waning days of the last Ice Age never really stopped. Historians of technology chart a continuous increase in the ability of our ancestors to modify and manipulate their environment. The domestication of animals, which started before the agricultural revolution, continued. Primitive wood buildings were replaced by more permanent structures of stone. As time went on, one invention followed another. Humans produced wheeled vehicles, began to use metals— first bronze, then iron—learned to plow their fields, and began to live in cities. Throughout this long, slow progression, bit by bit, humans were removing themselves from the natural world. And as they did so, both their numbers and their geographic spread increased. Instead of being just one more minor group of primates wandering around the plains of Africa, our ancestors established a position of dominance that we still hold.

I am aware that it has become fashionable in certain academic and environmental circles to decry this history, to shed fashionable tears over human arrogance and hubris, and to long (in a fashionable way, of course) for a simpler life that takes us back to our "natural" roots. Well, I beg to dissent from the fashionable view. To me, humanity's long struggle to build a life free from the dangers and constraints of the natural world is something to be celebrated, savored. I have enormous respect for those unsung men and women who first planted olive trees, learned to forge iron tools, and who (to quote Jonathan Swift) were brave enough to eat the first oyster. It is only because of their efforts that we can sit in comfortable coffee shops and debate the future of the relationship between humans and nature.

Even periods that are often considered backward and primitive—the Middle Ages in Europe, for example—exhibit evidence of substantial technological progress. Despite the fact that during this period the centers of science and learning lay in the Middle and Far East, Europeans reinvented the moldboard plow (opening the heavy soils of northern Europe to cultivation), developed the first primitive clocks, and started the manufacturing of soap.

The slow, steady progress I've described, punctuated here and there with important advances, continued substantially unabated until about five hundred years ago. At that point, a series of events in western Europe changed the face of the planet forever. These events are so important that we customarily denote them with capital letters—the Renaissance, the Enlightenment, the Scientific Revolution, the Industrial Revolution. Taken together, they mark a period of several hundred years during which the human ability to manipulate our environment and insulate ourselves from the process of natural selection went through a tremendous change. To my way of thinking, they mark the completion of the step that began when those first hunter-gatherers started helping plants that were useful to them prosper. In fact, we can imagine that the first step, which began in the Middle East ten thousand years ago, came to fruition with all of the capitalized processes listed above.

As with the agricultural "revolution," the completion of the first step was a gradual process, and it's hard to pick one event and say, "Here's where it all happened." In one sense, what took place in post-Renaissance Europe can be thought of as a continuation of the steady buildup of technical knowledge, inventions, and improvements that had preceded it. On the other hand, in any continuum, there is a point where "more" finally becomes "different." To me, the nineteenth century industrialized world, with its crisscrossing railroads, gas lamps, and steamships, was so different from medieval Europe that all talk of gradual change became meaningless. In terms of the metaphor I used at the beginning of this chapter, somewhere during this period, the sidewalk stopped its gradual rise and pitched sharply upward.

Look at it this way: Julius Caesar, Hannibal, and Napoléon Bonaparte all moved their armies in basically the same way—by having their soldiers walk. Yet only a half century after Napoléon, Civil War generals like Ulysses S. Grant and Robert E. Lee could move their troops by railroad. Somewhere in that time frame, a fundamental change had taken place—a change that was every bit as profound as the invention of agriculture.

As a physicist, I look at this change in terms of a single concept—energy. The energy supplied by muscles, whether they be in horse, human, or elephant, comes from the "burning" of carbohydrates, mainly a sugar known as glucose. In one way or another, this molecule powers every living thing on our planet. With the industrial revolution, we acquired the ability to substitute stored solar energy, in the form of coal burned in a steam engine, for glucose burned in muscles. It's that simple.

Throughout history, human beings have made some use of energy sources other than muscle, of course. Waterwheels have been used since ancient times, and wind power, most picturesquely represented by the windmills of the Aegean and the Low Countries, supplied energy. I will never forget, for example, a visit I made to Zaanse Schans, a national historical preserve near Amsterdam. This was the place where the Dutch merchant ships arriving with their cargoes of spices from what is now Indonesia would dock, and a row of windmills was erected to grind the spices before they were offered for sale. The buildings were solid, made of oak beams a foot thick, but when the wind was blowing, you could stand on a balcony and feel everything shaking with the movement of those great sails. There was plenty of power in that wind!

Nevertheless, both the waterwheel and the windmill suffered from a serious technical flaw. Both produced power that could only be used in situ. If you wanted to grind your grain or your spices, you had to take them to the place where the water flowed or the wind blew. Aside from those few fortunate spots, it was back to muscle power for everyone.

But another source was all around our ancestors, if only they had known how to unleash it. For hundreds of millions of years, sunlight had been pouring down on our planet, carrying a precious gift of energy. This energy was first stored in plant tissue and then, when the plants died, slowly converted into coal. This energy stayed underground, waiting patiently for someone to figure out how to tap it. In the eighteenth century, ten millennia after those first farmers settled down, the energy was unlocked by a group of engineers and tinkerers in England.

An eighteenth-century hardware merchant (what the English call an ironmonger) in western England by the name of Thomas Newcomen produced one of the first usable steam engines. A cumbersome affair, about the size of a small house, and, by modern standards, pretty crude, the engine developed less power than the average modern lawnmower. It worked by burning coal to boil water. The resulting steam was used to lift a piston in a cylinder, after which jets of cold water condensed the steam, creating a partial vacuum. The pressure of the atmosphere then drove the piston down. For the first time humans had a way of developing energy without muscle power and using that energy wherever they wanted.

In 1776, the year of the American Revolution, a Scottish engineer and entrepreneur named James Watt turned the clumsy, cumbersome Newcomen engine into a truly practical device. The details of his modifications are too technical (and, frankly, too boring) to go into here, but when he and Matthew Boulton started building steam engines at the Soho Works in Birmingham and selling them to miners in western England, the industrial revolution could fairly be said to be under way.

The fact that the first customers for the early steam engine were miners leads to an interesting bit of historical trivia. Whenever a hole is sunk in the ground, especially in a climate like that of Europe, it will start to fill with water from underground streams. From time immemorial, miners have battled with this brute fact of nature. In eighteenth-century England, the state-of-the-art drainage technology involved lowering horses (called pit ponies) into the mine and using them to turn wheels that lifted an endless line of buckets to the surface. To sell his engines to hardheaded mining engineers, Watt had to find a way to relate his work to their drainage problem. Consequently, he tested a few healthy horses, determining just how much work a horse could do in a typical work situation. An engine that could do the same amount of work was called a one-horsepower engine, because it would replace one horse in the mine. And even though most of us don't have to worry about draining mines, we still drive cars and use hand tools and motors rated in horsepower, courtesy of James Watt.

In any case, Watt's engine opened a new source of energy for the human race, every bit as important to the newly industrializing countries of

Europe as the acquisition of gold from the New World had been to Spain a few centuries earlier. In short order, smart engineers learned how to make the engine small enough to use it to power railway engines and ships, much as their descendants 150 years later learned to miniaturize the nuclear reactor so that it could power submarines. Using the stored energy in coal, nations around the world built railroads to bring people and goods to and from new factories in burgeoning cities. As with the period following the development of agriculture, the world changed forever.

Today, our first step is almost over. Humans no longer have to be content with what nature offers, but can grow their own food, produce their own energy, and create their own goods. The great wealth created by the industrial revolution drove technological changes so profound that, for the first time, the planet is home to a species that is no longer bound by natural selection. To put it as bluntly as possible, your chance of success in the traditional Darwinian game is no longer dependent on your genes. Whether or not those genes will make it into the next generation does not depend on your individual ability to hunt for food or defend your children against danger. We have, instead, built a technological society so strong and so pervasive that the Darwinian cycle of reproduction and selection is simply irrelevant to us.

When I was nineteen, I was struck by an attack of appendicitis while hitchhiking with my brother in Europe. After a quick visit to a doctor, I was whisked to a hospital in Amsterdam, where an emergency appendectomy was performed. A few days later, I walked out of the hospital and got on with my life. This rather mundane sort of experience—repeated in one form or another every day—depends critically on human technology and knowledge. Sterile surgery with anesthesia is a relatively modern development, and had I lived a few hundred years earlier, I probably would have died. And frankly, in the Darwinian sense, I *should* have died, because whatever there is in my genes that led to that attack should not be allowed to go to the next generation. Had natural selection been in control, none of my children would ever have been born.

But natural selection wasn't in control. Thanks to a long line of human inventors and scientists—a line stretching back at least ten thousand years to those early protofarmers—the attack of appendicitis was little more than a temporary inconvenience for me. In just the same way, diseases that used to decimate populations—diphtheria, scarlet fever, yellow fever, and many more—are routinely dealt with by immunization. Natural selection no longer weeds out those with weak immune systems, simply because a large part of our defenses against disease is supplied by our technology rather than our genes.

In fact, you can say that for a person living in a modern industrialized society, nature has almost ceased to matter. Of course, we depend on nature (heavily modified by human engineering) for basics like food and water, but by and large the nonurban "natural" world doesn't impinge on our lives. I look forward to the day when more of our fellow humans will be able to share in the fruits of the first step that are now concentrated in the developed world. For those of us who are fortunate enough to enjoy these benfits, however, aside from the occasional major storm, which usually is little more than an inconvenience, nature plays a very minor role in our lives.

This is good news!

For obvious reasons, I am very happy that humans no longer are bound by the laws of natural selection. Furthermore, I *like* living in a climate-controlled house, not having to care too much what the weather is like outside. I *like* not having to worry that a flood or a hailstorm will destroy my only source of food, placing my life and the lives of my family at risk. I *like* living in a society where my children's ear infections were treated with antibiotics, so that they are still alive and can still hear.

In fact, I would argue that the near completion of the first step is a process that should be celebrated as a great triumph of the human mind. Because of it, our planet, and our future, have been changed forever.

By understanding the beginnings of agriculture and the completion of the first step, we understand how we have arrived where we are. The obvious next question, of course, is where we are going. My argument is that we are already embarked on a second step, one that will make humans managers of the planet, and that the vague outlines of that step are already visible. Before we move on to this issue, however, we need to look at the world we live in and, in particular, to examine the way that the world is commonly portrayed. Only if we have a clear picture of how the world operates will we be able to have a meaningful discussion about how it is to be managed.

II

The Myths of Pop Ecology

4

Lost Eden

The technologies of the second step will give us the tools we need to become managers of the planet. In this sort of situation, however, tools aren't enough—we have to have a clear idea of how they are to be used. The first step in the management process, in fact, has to be an accurate picture of what that planet is really like.

Unfortunately, the picture of the Earth that exists in the popular media—what I call "pop ecology"—has at best a tenuous relationship to the actual place where we live. Some of this disparity comes from simple factual error, from the fact that it takes a while for good science to permeate into the public consciousness, so popular views often represent outmoded science. If this were all that mattered, it would be a fairly simple matter to find a remedy—some change in school curricula, some good TV series, some training sessions for media people, and we would be on our way.

My conversations with many friends about this book, however, have convinced me that people's attitude toward the environment (including my own) are often rooted much more deeply, in that shadowy world in each of us where fact, myth, and emotions come together to form our most strongly held convictions. Because of this, I would like to start my look at pop ecology by looking at that world.

Myths are powerful.

I suppose that many people reading a statement like this from a physicist imagine that it must denote a mood of sarcasm, or possibly dejection, or, at the very least, that it must be accompanied by a slight curl to the lip. After all, what two realms of human thought could be further apart than the world of myth and the hard, crystalline rationality of physics? Well, I am here to tell you that there is absolutely nothing negative implied in that sentence. Because I am a scientist, I am acutely aware of the limitations of my craft, and I am well aware that there are truths in this world other than scientific ones.

It is a minor linguistic tragedy, in fact, that the term "myth" has come to be synonymous with "false" or "charming but misguided." In its original Greek form of "mythos," myths conveyed powerful truths about the world, often in the form of allegories or stories. I would include religious and spiritual truths in this same category of mythos. Their goal is not to explain the working of the world, but to provide guidance for life. To paraphrase the late Stephen Jay Gould, myths tell us about the Rock of Ages, science about the ages of rock.

Myth and Science

The standard way of thinking about these two great forms of human knowledge—mythos and science—is to see them as distinct and separate. Gould, for example, referred to them as "non-overlapping magisteria." When scientists and religious leaders share the platform at conferences, as they often do these days, it has become customary for each to make a ritual bow of respect to the other's special sphere of influence before getting on with the subject at hand. And yet, as we shall see below, it is not uncommon for members of one "magisterium" to push over the line that supposedly separates the two. When this happens, "mythos" gets converted into "myth" in the modern sense.

This is important, because the confusion between the central truth of mythos and the results of these sorts of transgressions provides powerful emotional energy for many political debates in this country. Think, for a moment, about the push to teach creationism in the schools or some of the debate over the abortion issue. At bottom, these are not scientific debates (although there are elements of science within them), but debates in which people are advancing different variations of their deeply felt versions of mythos. I will argue that much of what I call pop ecology results from exactly the same form of transgression.

There are, in fact, two central myths (in the connotation of mythos) that have relevance to the way that human beings (or at least Westerners) perceive nature. These are the myth of the Garden of Eden and the myth of Noah's flood. Each of these contains an important central truth, and each can serve as the basis for transgressions of mythos into the world of science.

In the creation epic recounted in the Book of Genesis, we learn that after the creation itself, "The Lord God planted a garden eastward in Eden, and there he put the man whom he had formed" (Gen. 2:8).

We all know what happened next. Adam and Eve lived happily in their perfect garden until Satan, in the form of the serpent, convinced Eve to break the one commandment God had given and eat of the "tree of the knowledge of good and evil." As a result, they were cast out from the garden. Because of his disobedience, God tells Adam, "In the sweat of thy face shalt thou eat bread, till thou return unto the ground . . . for dust thou are, and unto dust shalt thou return" (Gen. 3:19). (Adam's lame excuse for his lapse—"The woman whom thou gavest to be with me, she gave me of the tree"—is perhaps best glossed over in this age of gender sensitivity.)

In Christian theology, Adam and Eve's decision to disobey the commandment of God becomes the Original Sin, transmitted from one generation to the next. In Catholic theology, it is transmitted at the moment of conception, while in some Protestant theologies it is transmitted through the act of sexual intercourse. (I got an interesting viewpoint on Original Sin when I asked a colleague about what was taught in her church. "Methodists believe in Original Sin," she sniffed, "but we don't go on about it.") In this scheme of things, Adam's mistake becomes a burden that each of us must carry. (This doctrine, incidentally, explains why it is so important to Christians that Jesus Christ was not born through the usual human process, but through a virgin birth.) The central truth of the Garden of Eden myth, then, can be characterized as a fall from grace, an irretrievable loss of a perfect past.

Nontheologians have, of course, made their own interpretations of the myth. Many commentators speak of the Garden of Eden story as representing a loss of innocence (especially sexual innocence). When the myth is applied to ecology, however, it takes on a particularly interesting form. It tells us that there was a time when humans lived in harmony with nature, when they lived lightly on the Earth. In many versions of this myth, the equivalent of the apple was the discovery of technology, of the means by which humans came to dominate nature. The central notion of the Garden of Eden myth—that things used to be better and have become worse because of human pride and arrogance—is easy to find in modern ecological writings.

I should note that the myth of a golden age is not confined to the Judeo-Christian tradition. In the *Republic,* for example, Socrates describes a vegetarian society in which people "reclined on rustic beds strewn with byrony and myrtle, drinking of their wine, garlanded and singing hymns to the gods in pleasant fellowship, not begetting offspring beyond their means." Similarly, in the *Statesman,* Plato talks of the long-ago reign of Cronos as one in which "savagery was nowhere to be found, nor preying of creature upon creature. . . . They had fruits without stint from trees and bushes; these needed no cultivation but sprang of themselves out of the ground without man's toil."

The Garden of Eden myth holds a strong grip on many modern urban intellectuals. I will never forget, for example, one of the most powerful environmental television ads of the 1970s, an ad that you still see replayed in specials devoted to the history of advertising. It opened with a shot of a birchbark canoe being propelled through the water by a buckskin-clad American Indian (played by a Native American actor with the marvelous name of Iron Eyes Cody). As he moved along, you began to see discordant elements entering the picture—a plastic bottle floating by, along with other bits of debris. As the camera panned back, you saw the canoe against the backdrop of a large riverside industrial plant. Finally, the camera closed in on the man's face, showing a single tear running down his cheek.

Many authors have cited this ad, together with the NASA photograph of the Earth floating in space, as one of the founding images of the contemporary environmental movement. It's easy to dissect this ad—most Native Americans I know, for example, use aluminum boats powered by outboard motors rather than birchbark canoes, and you have to wonder how many people who responded emotionally to Cody's buckskin shirt have ever seen a deer killed and skinned. But these are cheap shots, and they miss the point of the ad. Its purpose is not to portray a literal truth, but to invoke the power of the Garden of Eden mythos. And that it does beautifully. "Look," it seems to be telling us, "we once lived in a beautiful world, in tune with nature, but we fell from grace, and now we live in this . . . garbage." Pick up any modern writing about the environment, and you will find similar images.

The myth of the Garden of Eden (or, alternatively, the golden age) has a moral and pedagogical component. If, during the golden age, people practiced particular virtues (being vegetarians, for example, or not driving SUVs), then we, in this fallen age, can come closer to those ideal times by practicing those same virtues. A lot of the energy that drives the exhortations we hear to live a simpler life comes from the idea that by doing so we can emulate the Garden of Eden.

The second great myth that shapes our attitudes toward the environment comes from the story of Noah's flood (at least in the Western tradition). As recounted in the sixth chapter of the Book of Genesis, the flood was set into motion by human failings:

> *And God saw that the wickedness of man was great in the*
> *earth, and that every imagination of the thoughts of his heart*
> *was only evil continually (Gen. 6:5).*

As the story unfolds, God finds Noah, one righteous man, and tells him to build an ark to shelter some of earth's living things from the coming deluge:

> *And I will cause it to rain upon the earth forty days and forty*
> *nights; and every living substance that I have made will I*
> *destroy from off the face of the earth (Gen. 7:4).*

The flood waters came, and the Earth was inundated for 150 days, killing every living thing on the land. Then the waters receded, the ark landed on Mount Ararat (in what is now Turkey), and the land was repopulated. In the biblical story, God sets a rainbow in the sky as a promise that he will never again send a flood.

As with the Garden of Eden myth, there have been many interpretations of the flood. Scholars have noted, for example, that flood stories exist in almost all cultures, and they have argued that each of these stories chronicles a local event. After all, they say, floods occur almost everywhere on the planet and are usually traumatic for the local populations. It should be no surprise that they are preserved in tribal memory.

More recently, two geologists, William Ryan and Walter Pitman of Lamont-Doherty Earth Observatory in New York, have put together rather convincing evidence that conditions at the end of the last Ice Age were such that a massive flooding of the Black Sea Basin occurred about seventy-five hundred years ago. In their scenario, waters from the Mediterranean poured through the Bosporus (past the present location of Istanbul) and inundated what had been fertile farmland along the northern shore of that body of water. Their calculations suggest a situation in which the sea encroached on the land at the rate of about a mile a day. Small wonder people enshrined that event in myth! Ryan and Pitman suggest that it was this singular event that gave rise to the flood legends in much of Eurasia, as people displaced by the flood spread to other parts of the area (although, to be fair, this part of their thesis enjoys less support in the scientific community than their reconstruction of the flood itself).

But as always, scientific questions about the myth are different from the central mythical truths involved. Whereas the central truth of the Garden of Eden myth involved a fall from grace, the loss of a previous perfect state, the central truth of the flood story involves sin, punishment, and eventual redemption. In both cases, the cause of humanity's suffering is human action, but there is a subtle difference between the two myths. At the risk of oversimplifying, the message of the Garden of Eden myth is "Things were once perfect, but because of human misconduct, that state is lost forever," while the message of Noah's flood is "Humans have been bad and are going to be punished."

Just as the Garden of Eden myth, when applied to ecology, leads to the notion of the golden age, the myth of Noah's flood has its own ecological reverberations. In essence, it says that when human beings act in ways that are out of tune with nature (as they always seem to do), then ecological punishment is sure to follow. For example, when the *New York Times* editorialized about the Klamath Basin situation that I discussed in chapter 1, it jumped to the conclusion that "When an entire species is sufficiently threatened to require protection, it usually means that the same ecosystem will fail humans who depend on it as well." As with Noah's flood, human misbehavior (taking too much water for irrigation) will lead to retribution (the failure of the ecosystem). My sense is that a great deal of the unfortunate scolding and carping tone of modern environmental writing derives its energy from this particular aspect of the myth of Noah's flood.

Transgressions

The world of mythos has one set of truths, then, and the world of science has another. Strictly speaking, there is no way that the truths of one can be verified by the methods of the other. To ask for a "scientific" proof of the Garden of Eden myth, for example, would require a scientific instrument for detecting a "state of grace"—something that is preposterous on its face. In the same way, questions about geological events in the Black Sea Basin seventy-five hundred years ago cannot be answered by retelling the flood myth, but only by examining geological evidence through the methods of science. This is the basis for the notion of the two worlds—mythos and science—as nonoverlapping magisteria.

In general, the line between the two magisteria is pretty clear. If a question involves knowledge about the physical universe ("How old is the earth?" or "Why is the sky blue?"), then it falls within the realm of science. If, on the other hand, it involves questions of morals or values ("What is the good life?" or "Why are we here?"), then it falls within the realm of religion or mythos. What I am calling a transgression occurs when someone takes a truth from one magisterium and asserts that it is also a truth in the other, without admitting that the criteria for truthfulness in the new setting are different from those in the old.

Let me give an example that keeps popping up in American education—the attempt to impose the teaching of special creation on the biology curriculum as a substitute for (or alternative to) the theory of evolution. Although I have found this trend in followers of many religions, in the United States it seems to be concentrated among a small group of fundamentalist Protestant churches, so I will approach the discussion from that point of view. (I should add that in my youth I belonged to one of these churches and lived the complete evangelical life, including a stint as a street corner preacher in Chicago. Consequently, I can speak about this belief system from experience.)

The central truth of evangelical Protestantism involves the doctrine of salvation through the sacrifice of Christ. In some churches, this is coupled with a fervent belief in biblical inerrancy—the doctrine that the Bible contains not just moral truths, but scientific ones as well. In a still smaller number of churches, this belief becomes a doctrine that holds that the words in the Bible must be taken literally and cannot be interpreted as metaphors or literary flourishes. In this view, when the Book of Genesis speaks of God creating the world in seven "days," the words must be taken literally. This belief is usually coupled with the statement that the Earth is only about ten thousand years old, a number arrived at in the eighteenth century through a dubious interpretation of various Bible passages.

This progression of statements goes from the doctrine of salvation by faith (which is unmistakably in the realm of mythos) to statements about the age of the Earth and the nature of living things (which are unmistakably in the realm of science). It is, in other words, a classic case of transgression between magisteria.

To see this, ask yourself how you would go about establishing the truth of the statements at the two ends of the chain. For the religious statements, you would use theological arguments of the sort taught at seminaries. For the statements about evolution and the age of the Earth, on the other hand, you would talk about the occurrence of fossils and the overlap of DNA among different species, described in chapter 2. Once the chain of argument pushes over the dividing line between the magisteria, the criteria for establishing truth change. The basic fallacy of creationism in its many forms and guises is that it insists on applying religious criteria to scientific statements, which just doesn't work.

When I started writing this section, I tried very hard to come up with an example of a transgression that went in the other direction—an incursion that went from science to religion. There are a few of these, although, as with the religious incursions, they are confined to a relatively small number of people. A few writers (the British biologist Richard Dawkins comes to mind) have stated that evolution proves that there is no purpose to life, and theoretical physicist Steven Weinberg closed his classic book *The First Three Minutes* with the statement: "The more the universe seems comprehensible, the more it seems pointless." (To be fair, Weinberg later stated publicly that he wishes he'd never written that sentence.) Some religious thinkers see these statements (rightly, I think) as incursions into their special domain. But aside from a few instances like these, I can't think of any major examples of such incursions. I suspect modern scientists are so busy pushing back the frontiers of knowledge that they seldom think about questions of religion at all.

The view of the world that grows out of the Garden of Eden and Noah's flood myths is pretty gloomy. In his marvelous book *The Skeptical Environmentalist*, the Danish statistician Bjørn Lomborg calls writing about pop ecology "The Litany," which represents a summary of the views of pop ecology. His version of The Litany goes like this:

> The environment is in poor shape. . . . Our resources are running out. The population is ever growing, leaving less and less to eat. The air and water are becoming more polluted. The planet's species are becoming extinct in vast numbers. . . . We are defiling the Earth, the fertile topsoil is disappearing, we are paving over nature, destroying the wilderness, decimating the biosphere, and will end up killing ourselves in the process.

For completeness, I should point out that there continues to be intense criticism of Lomborg's book—much of it politically inspired and unwarranted, in my opinion—but that doesn't change the fact that he certainly got The Litany right.

I will argue that propositions like those in The Litany are related to the myths (in the sense of mythos) we discussed above, but represent transgressions from the mythical into the scientific world. Like creationism, they are the end product of a chain of reasoning that starts in the one world and then slips across the boundary into the other, without any recognition that the criteria for establishing truth have changed.

The Birth of Pop Ecology

I define "pop ecology" as the kind of ecology presented in the popular press and in school textbooks, and, unfortunately, embodied in The Litany. It tends to be made up of ideas that were once mainstream science, but have been abandoned or modified significantly over the years. In other sections of the book, I will deal with many of them in detail, but I will state them and identify their mythical origins here. I know of no scientist who would defend all of the following propositions, and few who would defend any of them in their most radical form. Nevertheless, they describe the world as it exists in the minds of all too many members of the general public.

- The balance of nature: this myth holds that, left to itself, nature will tend toward a state of stable equilibrium, with all species in a finely tuned balance with each other. This is an outgrowth of the Garden of Eden myth, an idea that before technology there was a golden age of balance and harmony.
- The noble savage: another outgrowth of the Garden of Eden myth, this is the notion that primitive people have discovered the secret of living in harmony with nature, a secret lost to modern technological society.
- The poisoned planet myth: the belief that humans are flooding the planet with chemicals against which living systems have no defenses and that will therefore cause all manner of harm. It's a complicated myth, because it's partly true and partly false. It is, however, a clear example of the sin-and-retribution theme associated with Noah's flood.
- The stable climate myth: the idea that, left to itself, the earth's climate will be inherently stable (at least over reasonably long timescales) and that only human intervention changes it drastically. This is another Noah's flood myth, with human misconduct being punished by ecological disasters like global warming.
- The extinction myth: the oft-repeated proposition that human activity is causing an extinction of species that the planet has not seen since the end of the dinosaurs. Again, this myth is complicated because extinctions are undoubtedly occurring,

although probably not at the level posited. This is another retribution myth like Noah's flood.

In addition to these myths, which are based firmly in the mythos of the Garden of Eden and Noah's flood, there is another myth of a more technical nature, having to do with more modern understandings of ecosystems:

- The complexity myth: the idea that natural systems are so complicated that it is impossible to predict the consequences of any human intervention. The upshot of this notion is that humans should never attempt to change the world, but should allow natural processes to go on unhindered. The real question, of course, is just how complex ecosystems are and exactly how well they can be managed. In its mythical sense, however, this argument is usually invoked to forestall human interference with natural process by invoking the threat of the sin of pride and inevitable retribution—the Noah myth.

If you believe that the world we live in is governed by the myths of pop ecology, then that world is a pretty awful place, full of angst, sorrow, and blame. Each of the myths carries a sense of loss or a sense of transgression against nature. Fortunately, however, they don't really describe the world we live in, but the only way to see that is to examine the way that world actually functions.

5

The Myth of Stability

During the twelfth century, the government of France faced a small but serious problem. Wines from a neighboring country had become so popular, and imports so extensive, that serious consideration had to be given to negotiating trade agreements to protect French vineyards.

Now this may not seem like surprising news. After all, France is surrounded by wine-making regions—Germany, Spain, Italy, and Switzerland. What is interesting is the source of the threat to French viniculture. The wines in question came from England—cold, cloudy, "A Foggy Day in London Town" England! (Fellow aficionados of John Mortimer's *Rumpole of the Bailey* series may rejoice to know that our hero's "Château Fleet Street" has a basis in historical fact.)

When we think about our planet, we often tend to make what I call the Fallacy of the Snapshot. We assume that the world has always been exactly as it was when we first became aware of it. In this case, it is tempting to assume that since England is cold and rainy now, it must have always been cold and rainy. Yet, as this little-known episode from agricultural history shows, only a relatively short time ago the climate of that island was quite different from what it is now. In point of fact, during this period, vineyards in Europe were found up to three hundred miles north of their present northernmost locations.

The vanished vineyards of England teach us something important about our planet: the climate has always been changing. In fact, in this chapter, I will sketch some past climate changes—all of them unrelated to human activity—that make recent changes like the one described above look positively picayune.

And this, in turn, will lead us to consider one of the prevailing myths of pop ecology—the myth of stability. This may well be the most pervasive myth promulgated about the nature of our planet. The story line of the myth goes something like this: our planet and the ecosystems that exist on it are a finely tuned system. This system, through the action of nature, has achieved an exquisite and stable balance, with each part dependent on all the others. The interlocking web is perfect and delicate, like a ball balanced on a knife edge. Once this equilibrium has been reached, it remains unchanged—after all, how can you improve on perfection? Disturb one tiny part of this edifice, the myth says, and the whole thing could come crashing down like a house of cards.

The motto that comes from the myth could be characterized as "Don't rock the boat" or, in the more elegant phraseology of the current environmental debate, as the "precautionary principle." Because anything you do could disturb the finely tuned balance of the planet, the best policy is not to do anything until you can prove beyond a shadow of a doubt that it will cause no problems. Clearly, if the world is as delicately balanced as that, the project of managing the planet is going to be very difficult, if not impossible.

A moment's reflection on the nature of the evolutionary process, as described in chapter 2, should raise immediate questions in your mind about the myth of stability, however. Natural selection is a slow process whose net effect is to produce individuals that can survive in a given environment. As time goes by, succeeding generations get better and better at exploiting their ecological niches. *Provided* that those niches don't change, and *provided* that the environment stays the same long enough for natural selection to work its magic, you can imagine a balanced, finely tuned system evolving.

But that's where the story of the English vineyards comes in. No matter where you look in the history of our planet, you find change. Climates are not steady, but locked in an eternal flux. Because of this fact, ecosystems are always changing, always playing catch-up, always dealing with some new feature of the environment, be it a volcanic eruption or a new ice age. "Stable" is simply not a word that describes the world we live in. That's the beauty of life on our planet. It's tough, it's resilient, it survives.

The English vineyards, in fact, were a product of what is called the Medieval Warm Period, a centuries-long period of mild weather that produced, among other things, the wealth that created the great Gothic cathedrals of Europe. Lasting roughly from 900 to 1300 A.D., the Medieval Warm Period was followed by a colder period called the Little Ice Age, which ended (roughly) in the middle of the nineteenth century.

Given this picture, we can see that the problem with the snapshot is that what we see depends on when we take the photograph. In this chapter, I want to talk about some climate episodes in the Earth's past to give you a sense of just how mercurial the climate is, with some specific examples of what sorts of climates we might have considered "normal" had we decided to take our snapshot at a different time.

Measuring Past Climates

As it happens, a great deal of scientific work is required to make authoritative, data-based statements about the past climate of the Earth, so it would probably be a good idea to pause for a moment and think about how we can know what things used to be like.

Even in historic times, climate records are hard to come by. Remember that the first thermometer wasn't built until the seventeenth century, so before then it would have been impossible for anyone to maintain a temperature record. Over the last couple of centuries, we have weather records for only a few cities in Europe and North America. Even these records are somewhat idiosyncratic, since they were often kept by private individuals—one of the best eighteenth-century climate records, for example, was a daily log kept by Thomas Jefferson at his home at Monticello in central Virginia. Another weather-keeping tradition that continues to this day can be seen at the meetings of the Philosophical Society of Washington. Founded in the mid-nineteenth century and modeled after the society begun by Benjamin Franklin in Philadelphia, the minutes of each meeting close with a statement of temperature and weather conditions. Someday, I suppose, someone will look at these records and incorporate them into future climate models.

But because of the unavailability of direct data, when scientists want to talk about climates from more than a few centuries ago or recent climates in places other than Europe and the east coast of North America, they use what are called proxies. As the name implies, a proxy is a measurable quantity that can be used to estimate or, in some cases, calculate things like temperature and rainfall.

For example, scientists have used the price of wheat on the London grain exchange as a proxy for summer temperature and rainfall in Europe, the idea being that warm summers with abundant rainfall will produce bumper crops and, therefore, lower prices. In the same way, going back to medieval tax records for Alpine villages can tell us something about the advance and retreat of mountain glaciers, since fields buried under ice would be removed from the rolls, then reentered when the glacier retreated. The freezing date of lakes in Japan, the number of weeks each year that the coast of Iceland was icebound, the frequency of dust storms in northern China, and the date of the blooming of cherry trees in Kyoto, Japan, have all been used as proxies for climate.

For my money, the cleverest use of a proxy for deducing weather patterns was made by the German scholar Hans Neuberger in 1970. He looked at sixty-five hundred paintings that had been done between 1400 and 1967. Instead of looking at the main subject of the paintings, however, he looked at the clouds that the artists painted in the background. Arguing that artists would have given little thought to this part of the painting, but would put in what they considered to be typical, he documented that the incidence of low clouds in the background increased sharply after 1550, then fell slightly after 1850. The English painter John Constable, for example, regularly depicted over 75 percent cloud cover in his famous landscapes, much more than would be typical today. Like the vanished vineyards of England, these artistic records document significant changes in climate in recent historical times.

To go back before the existence of historical records requires a different sort of proxy, one based more on analytical techniques. In the early part of the twentieth century, for example, scientists in the American Southwest began studying annual rings in long-lived trees. As you likely know, trees add one ring each year, so that by counting inward from the present, you can tell when a particular ring was formed. Wide rings indicate a good growing year, with ample rainfall and warm temperatures, while narrow rings indicate environmental stress. In this sense, the tree "remembers" what the weather used to be like. Using cross sections bored from very old trees such as the bristlecone pines that grow on the eastern slopes of the Sierra Nevada mountains, as well as ring patterns from dead trees, scientists have been able to extend the climate record measured in this way back about ten thousand years.

Let me interject a personal note here. A number of years ago, I wrote an article on tree-ring dating for a national magazine and as a result had an opportunity to visit the Laboratory of Tree-Ring Research. It's located on the campus of the University of Arizona—under the stands of the football stadium, believe it or not! The experience was unforgettable. I walked from a typical athletic corridor, reeking of sweat and liniment, into a room dominated by the sweet, resinous smell of pinewood. There, scientists pored over their microscopes, examining cores that had been taken from the bristlecones. By studying material that had been formed in the past, they were able to reconstruct a climate long gone. It was work like this that led to the discovery, for example, of the twenty-two-year drought cycle that has plagued the American West for centuries.

To go further back than ten thousand years, scientists have to identify something else that was formed long ago and can "remember" what the weather was like when it formed. The glaciers and ice caps of Antarctica and Greenland, for example, provide some of the best proxies we have for long-term weather. When winter snow falls on the ice sheet, it packs down. During the summer, the upper layer of the snow melts a bit and compacts even more, and may incorporate some windblown dust as well. The result: a dark streak in the ice, analogous to the dark layer in tree rings. By drilling into the ice and counting annual layers, scientists can locate ice that was formed hundreds of thousands of years ago.

The poet Frauçois Villon once asked, "Where are the snows of yester-year?" We can now answer confidently, "In the Greenland ice cap!"

There are two important kinds of "memory" in this ice. When the snow falls and starts to compact, tiny air bubbles are formed in the ice. Thus, little pieces of long-gone atmosphere survive today, waiting to be studied. By analyzing the air in these bubbles, scientists can determine (for example) how much the amount of carbon dioxide in the atmosphere has varied in the past.

Another important indicator is the oxygen in the ice itself. Like all chemical elements, oxygen comes in several varieties, called isotopes. The most common form of oxygen has eight protons and eight neutrons in its nucleus and is called oxygen-16 to indicate this fact. The most common other form has eight protons and ten neutrons and is called oxygen-18. It is identical in every way to oxygen-16 except that the extra neutrons give it a little more heft. Roughly one water molecule in five hundred will be made from oxygen-18 and hence will weigh a little more than its neighbors.

When the temperature goes up, more water will evaporate from the oceans. The lighter water molecules made from oxygen-16 will be more likely to go into the atmosphere, form clouds, and return to the surface as snow. Thus, the ice that forms from this snow will have relatively less oxygen-18 than ice formed when the temperature is lower. By measuring the so-called isotope ratio—the relative amounts of the two oxygen isotopes—scientists can reconstruct the temperature of the earth as it was in the past. The Greenland ice cores can take us back over one hundred thousand years, and one core, taken near Vostok station in Antarctica, goes back a full four hundred thousand.

We can go even further back in time by using the other half of the process we've just described. If the temperature goes up and water with oxygen-16 leaves the ocean, then what is left behind will have relatively high levels of oxygen-18 in it. By examining sediments that formed on the ocean floor, we can use this information to go back many millions of years.

The bottom line of all this is that the Earth remembers what the climate used to be, and if we are clever we can get it to share those memories with us.

The Medieval Warm Period and the Little Ice Age

Have you ever wondered how Greenland got its name? For most modern Americans, our only contact with the place occurs when, for one reason or another, a transatlantic flight is routed north and the mysterious topography of the island passes under us. There certainly doesn't seem to be much about it that is green these days, particularly in the winter. Even in the summer, the best I've ever seen looked like some scrubby lichen and grass clinging to inhospitable cliffs. Yet around 980 A.D., hardy Norsemen not only gave the island its name, but built a flourishing settlement there. The Greenland settlement survived until about 1350—it was there roughly as long as European settlers have lived in North America.

Both the settlement of Greenland and its ultimate abandonment are symbolic of the changes in weather patterns over the past millennium. The Medieval Warm Period was marked by temperatures comparable to those of the latter part of the twentieth century and, more importantly, by long periods of climatic stability. For an agricultural society like Europe, always poised at the edge of disaster, where a few years of drought or flood could mean famine, it was a time of unequaled prosperity, a veritable golden age. As I mentioned above, the great Gothic cathedrals that dot the European landscape are a legacy of the prosperity enjoyed during that period.

But the climate always changes. Beginning around 1300, for reasons we don't really understand, that change started again. Perhaps there was a subtle shift in the Earth's orbit, or perhaps the sun itself dimmed by just a fraction. Whatever the cause, the good days were over in Europe. It wasn't so much that it got colder (although it did), but that the climate became erratic. Storms swept in, inundating coastal settlements. Bad harvests became more frequent. During the great cold snap of 1586 to 1595, winter temperatures in Europe were fully two degrees Celsius (four degrees Fahrenheit) below their twentieth-century values. And if this doesn't seem like a lot to you, remember that during the last Ice Age (see below), the temperatures were only about five degrees Celsius (ten degrees Fahrenheit) below their current values.

There are many ways to put a human face on the Little Ice Age. Dutch painters of the seventeenth century routinely depicted winter scenes with rivers and canals frozen over. In London, whole villages would be set up on the frozen Thames each winter, rather like the wintertime "villages" of ice-fishing shacks that appear on lakes in Minnesota. My own insight into the Little Ice Age comes from visits to museums in the Washington, D.C., area. Paintings from that era depict the heavy wool frock coats that men wore, summer and winter, in the eighteenth and early nineteenth centuries that would be impossible to wear today. They make sense only in the context of much cooler summers than today's Bermuda-shorts-and-open-neck-shirt weather.

But, as must all climate conditions on our planet, the Little Ice Age came to an end. Sometime in the middle of the nineteenth century, also for reasons we don't understand, global temperatures began to climb—the cause seems to be an uncertain mix of natural and anthropogenic causes. But the lesson of the last millennium is clear: no matter what the climate of the Earth is like, the one thing you can be sure of is that it will change.

Ice Ages and Heinrich Events

The climate fluctuations of the last thousand years, while significant, were really rather mild as these things go. Eighteen thousand years ago, periodic changes in the Earth's orbit and axis of rotation plunged the planet into an ice age, in which large glaciers moved toward the equator from the poles. In North America, the glaciers came down as far south as New York City and Chicago, creating places like Long Island when they melted and dumped the rock and gravel they had been carrying. These weren't the first glaciers to descend on the planet, nor will they be the last. In fact, climatologists reckon that we are living in what they call an interglacial—a term I find chilling in every sense of the word.

The basic theory of ice ages was worked out by the Serbian engineer and mathematician Milutin Milankovich in the early part of the twentieth century. Basically, he noted that because of the gravitational pull of the other planets, three subtle but distinct changes take place in both the Earth's orbit and the direction of the Earth's axis of rotation. These changes are slow—the fastest involves a timescale of twenty-six thousand years—but when they all act together, they induce a slow cooling that sends the glaciers toward the equator. When they pull in different directions (for example, when one pushes the planet toward cooling and the other two push it toward warming), the glaciers recede and the planet moves into an interglacial. All in all, the Vostok ice core shows three separate glacial periods in the last four hundred thousand years, each with about ten glacial advances and retreats.

But ice ages are an old story—they were discovered by scientists in the nineteenth century and were, in any case, relatively slow processes. A temperature change of five degrees Celsius (ten degrees Fahrenheit) took thousands of years to happen, and no one had any trouble imagining that delicately balanced ecosystems could migrate north and south, maintaining their complex web of interrelationships, on such a leisurely timescale. What is surprising were discoveries scientists made about climate changes *within* the last ice age.

One method of studying past climates is to send ships out on the ocean to take core samples of the bottom sediments. Scientists doing this in the North Atlantic soon uncovered a rather strange phenomenon. As they penetrated deeper and deeper into the sediments, they began finding layers of jumbled rocks that seemed out of place. Spaced every seven to twelve thousand years, these layers of rocks had a chemical composition that indicated that their place of origin was somewhere around Hudson Bay in Canada. Furthermore, temperature measurements indicated that these layers were correlated with enormous and rapid changes in temperatures. In fact, about the time the layers of rock were being laid down, the temperature increased by about five degrees Celsius (ten degrees Fahrenheit) in a matter of decades! The temperatures would stay elevated for a few thousand years, then fall equally precipitously back to their normal levels. Called Heinrich events, after Hartmut Heinrich, the scientist who first uncovered them, these events may well be the most dramatic and rapid changes in climate in the planet's history.

One of the first people to propose an explanation of these events was Douglas MacAyeal of the University of Chicago. Doug is a glaciologist, one of those strange breeds of men who actually enjoy camping out on the ice for months at a time while they study Antarctic glaciers. Doug's suggestion was simple. As the glacier built up over Canada, the ice sheet got thicker and heavier. Most of the rocks of northern Canada are strong, quite capable of bearing the weight of a three-mile-thick layer of ice. In the Hudson Bay area, however, the rocks are softer. As the overburden increased, they began to crumble and fracture. The powdered rock, mixed with meltwater from the ice, formed a gooey slurry—something with about the consistency of toothpaste. Slowly, the ice sheet began to slide out over this lubricated surface. As the ice sheet sloughed off, an armada of icebergs began sailing out into the North Atlantic, each carrying a cargo of rocks frozen into its structure. When the icebergs melted, those rocks fell to the ocean floor, to form layers that scientists would discover thousands of years later.

To understand why an event like this could produce drastic changes in the climate, you have to understand a few things about the properties of water and the role of the oceans in regulating climate. Water first. The basic idea is that the waters in the ocean are like the materials in a salad dressing; denser water will sink toward the bottom while less dense water will float to the top.

What makes water dense? A number of things have an effect. When water is heated, it expands, so, in general, warm water is less dense than cold water and will be found at the surface. (There is an interesting exception to this rule near the freezing point, but that doesn't concern us here.) In the same way, salty water is denser than fresh water, so salty water will tend to sink. The structure of the oceans results from a complex interplay of temperature and salinity.

The role of the oceans in determining the Earth's climate results from two simple facts: (1) more energy in the form of sunlight falls in the tropics than at the poles, and (2) the Atlantic Ocean is saltier than the Pacific. The great circulation patterns of the ocean are designed to counteract these effects.

The warmth of the tropics has to do with the tilt of the Earth's axis. The saltiness of the Atlantic requires a little more explanation. In the tropics, the prevailing winds (the so-called trade winds) blow from east to west. Month in and month out, dry winds from the Sahara blow to the west, evaporating water from the Atlantic, leaving the ocean saltier. Normally, this evaporation wouldn't matter, since the water would eventually return to the ocean as rain. In this case, however, a good fraction of that Atlantic water falls on the far side of the Isthmus of Panama. Thus, because of the trade winds, the Atlantic gets saltier while the Pacific is diluted, becoming less salty.

Incidentally, in keeping with our theme of noting changes in the Earth, this saltiness gradient between the two oceans is relatively recent, since the

Isthmus of Panama formed only about three million years ago. Before that, the two oceans were connected between the Americas.

In any case, the role of both the atmosphere and the ocean is to move heat from the tropics to the poles and salt from the Atlantic to the Pacific. Basically, it is this redistribution (especially the redistribution of heat) that gives rise to climate. Some of this process is fairly familiar. We know that the Gulf Stream, which carries warm water north along the western Atlantic, becomes the cold Canaries current as it returns to the tropics along the coast of Europe. In a recently discovered pattern that scientists have christened the Great Conveyor Belt, warm water moves north in the Atlantic to the region of Iceland, where it gives up its heat and sinks, flowing slowly southward as a cold bottom current. Rounding the Cape of Good Hope, this bottom current splits. One branch flows south of Australia and north into the central Pacific, the other north into the Indian Ocean. Both branches well up and flow eastward as surface currents, joining up again east of Africa before flowing back into the Atlantic.

The operation of the Great Conveyor Belt depends on water in the North Atlantic being able to sink. Once the iceberg armada has sailed, however, there will be a great pool of cold, fresh water floating on the surface of the North Atlantic, a pool that impedes the belt's flow. The idea, then, is that each Heinrich event corresponds to a slowing down of the Great Conveyor Belt, with a consequent abrupt change in climate. It takes a few thousand years for the pool of cold water to dissipate, after which the belt turns back on and the climate returns to normal. "It's as if there were a layer of foam over the North Atlantic," says MacAyeal. "The interaction between the ocean and the atmosphere is changed, and the Conveyor Belt dumps less salty water into the Pacific. It's like getting a weak flush in your toilet."

Since the initial discovery of the Great Conveyor Belt, scientists have learned that each Heinrich event is preceded by a series of smaller, weaker fluctuations in temperature. They used to be called flickers, but now go by the more scientific-sounding name of Dansgaard-Oeschger events. In addition, we now know that the story behind these changes has to do with more than the glaciers. It also involves changes in freshwater runoff from the continents, a phenomenon that increases in importance when glaciers melt.

But, in the end, the amazing thing about Heinrich events isn't the size of the temperature change associated with them, but the speed with which the changes took place. Temperature changes equivalent to moving Miami to Boston and back again have taken place in a matter of a few years! Compared to the relatively small and leisurely events of the past millennium, these were huge climatic disruptions. And yet, as far as we can tell, there was no accompanying extinction of life-forms.

Snowball Earth

And now for the biggest climate swing in our planet's history. Scientists studying rock formations as far apart as Namibia and Boston harbor have found evidence for a series of climatic events dating back to about six hundred million years ago, back when all life on our planet was in the form of single-celled organisms. For reasons we don't understand, the climate suddenly started to cool. Perhaps it was the aftermath of a meteorite collision raising a dust cloud that enshrouded the planet; perhaps the sun went through a temporary cooling phase. In any case, ice started to form at the poles and move downward. As more ice formed, more sunlight was reflected and the cooling intensified. Over thousands of years, the ice began to move out from the shorelines of the continents, eventually covering the entire northern part of the oceans. This, of course, shut down ocean circulation and added to the alteration of the climate. According to our calculations, when the ice reached a position about 30 degrees from the equator, a "tipping point" was reached, and the rest of the oceans froze over in a matter of a hundred years.

So there our planet sat, frozen solid, an ice-covered ball in space—a "Snowball Earth." The ice-shrouded land and oceans reflected almost all of the incoming sunlight, so there was no chance that the sun could save us from this situation. But our planet had more cards up its sleeve. Deep within its interior, far below the icy surface, radioactivity and heat boiled up, creating volcanic eruptions that easily melted their way into the atmosphere. Gradually, the volcanoes released gases like carbon dioxide into the air. Under normal circumstances, this carbon dioxide would be taken back into the earth, forming ocean sediments and weathering rocks. The layer of ice on the Snowball, however, cut the atmosphere off from the surface, so the carbon dioxide accumulated in the atmosphere, blocking outgoing radiation, wrapping an ever-thickening insulating blanket around the planet. Eventually, the resulting greenhouse effect raised the temperature at the surface so high that the ice began to melt. As soon as some surface near the equator was freed from its overburden, of course, the whole Snowball scenario ran in reverse. As the uncovered land and ocean absorbed incoming solar radiation, the planet's temperature rose, melting more ice and uncovering more land, raising the temperature even more. Within a hundred years, the ice melted and the Earth returned to its normal state as a planet with liquid water on its surface.

The Snowball Earth scenario involved changes of tens of degrees in planetary temperatures on timescales of hundreds of years—arguably the most dramatic climatic changes in our history. But life survived, probably in warm pools around volcanoes and hot springs. When the planet came back, life did, too. In fact, some scientists have suggested that the appearance of multicelled life (like us) was triggered by the Snowball events.

Maybe.

Could there be a Snowball in our future? I think the answer is probably no. It's not that I think the climate has suddenly become stable—it's just that another aspect of change in our solar system is that the sun has been growing steadily more luminous over the past four and a half billion years, as stars will. In fact, since the last Snowball event six hundred million years ago, the sun has become about 7 percent brighter than it was in those days. This means, I think, that our planet is no longer poised at the edge of a Snowball. But for all that, the Snowball episodes give you a sense of how variable the Earth's climate can be without any intervention from humans.

The Earth's climate is not a finely tuned, balanced system. Instead, it presents life with a constantly changing, constantly variable environment. And living things, in turn, have learned to adapt to change. Our planet is simply not a delicate thing, at the mercy of the first change in the climate.

This fact makes the task of planetary management both simple and more complex. On the one hand, it means that we don't have to worry that anything we do will "rock the boat," since this boat does plenty of rocking on its own. On the other hand, it means that there may be trigger points in the climate system that signal the onset of drastically new climate regimes and that an understanding of where those points are will be an important component of any future planetary management scheme.

6

The Myth of the Pristine Wilderness

The deserts of northern Peru are among the driest and most desolate places on the face of the planet. They're in the tropics, so the prevailing winds come from the east. Clouds are lifted up as they approach the Andes Mountains, dumping their rain to produce the verdant forests of the western Amazon, while clouds from the Pacific Ocean to the west of the desert are blocked by a range of coastal mountains. The result: a haunting, desolate landscape without a visible trace of vegetation or animal life.

The only exceptions are occasional valleys where rivers from the Andes run into the Pacific. These valleys were the sites of the pre-Inca civilization that flourished on the South American continent. In the late 1980s, in a small town called Sipan, archaeologists uncovered some unlooted graves whose funerary treasures rivaled those of King Tutankhamen in Egypt. Roughly contemporaneous with the Romans, the builders of those pyramids and tombs were known as the Moche (after a Peruvian river) and had built one of the great civilizations that rose and fell in the New World in the years before the Inca and Spanish conquests.

I was going to see those tombs, but as I came out of the arid desert into the broad valley that had nourished these long-vanished people something else entirely caught my attention. I noticed that in addition to the green region around the river, all through the valley there were green fields supporting a complex aquatic ecosystem. White ibises stalked frogs and fish in small waterways, surrounded by the sorts of reeds and fronds that you usually associate with swamps. I wondered how this ecosystem had grown up, given the fact that the current inhabitants of the area did not appear to have the resources to build an elaborate irrigation system. Then it struck me. What I was looking at were irrigation canals that had been built by the Moche empire, thousands of years before. Through all the slow centuries, water had been diverted into those ditches, keeping the fields green, supporting the people in their villages, and, incidentally, supplying those ibises with their lunches. If there ever was an example of human beings modifying and managing their environment, here it was right in front of me.

As we shall see, one of the great discoveries of modern archaeology is that the first people to come to the Americas carried out massive modifications of the environment in the years before European contact. This discovery undermines what could be called the founding myth of modern environmental philosophy—the myth of the noble savage and the myth of the pristine wilderness.

The myth goes something like this: before European contact, Native Americans lived in a state of communion with nature. They lived lightly on the land and cooperated with nature rather than trying to control it. This is the type of life that humans were meant to lead, a life from which, sadly, modern humanity has fallen in its pursuit of greed and materialism. As a result of the way of life adopted by Native Americans, the wilderness in the Western Hemisphere was in a pristine, untouched state when Europeans first saw it; and a return to that state should be the goal of any environmental or conservation policy.

This last statement is important, because it carries within it both an ideological and a political message. The ideological message, which is a variation on the Garden of Eden myth discussed in chapter 4, is that there once was a time when the relationship between human beings and nature was both very different from and far superior to what it is now. The political message is that the condition of the Western Hemisphere at the time of first European contact can serve as a template against which to measure the success of our stewardship of nature.

The noble savage myth has become so pervasive in American culture as to be almost invisible, but think about it: when was the last time you saw a movie in which a Native American was anything less than stoic, honorable, and noble? When was the last time, in other words, that you saw a Native American portrayed as a human being rather than as a stereotype? How many times have you heard that Native Americans never owned land, but believed that nature had to be shared? That they were good environmentalists, with a reverence for nature? How many times have you seen that quote attributed to Chief Seattle—the one that starts "How can you buy or sell the sky?"—on sale in tourist stores, despite the fact that it was written in 1970 for a movie? (The question of the authenticity of an older version of the speech, supposedly transcribed in 1855, is one of those murky historical puzzles that will probably never be resolved.)

This list could go on, but I think I have made my point. There is a pervasive belief in popular American culture that the people who occupied this continent before the arrival of Europeans knew how to interact with nature in a more wholesome and healthy way than we do. What I will argue in this chapter, and what recent advances in archaeology show us, is that Native Americans were no different from other human beings everywhere— they exploited and controlled their environments to the maximum extent permitted by their technologies. Furthermore, the growth of the noble savage and pristine wilderness myths were actually the result of a historical accident—a drastic drop in Native American populations caused by the unwitting introduction of new diseases by Europeans and their livestock.

Populating the Wilderness

Let's start with the first introduction of humans into what had been a human-free environment. The conventional wisdom has been that human beings walked into North America from western Asia about twelve thousand years ago. At that time, the retreating ice sheets of the last glacier opened up a land corridor hundreds of miles wide at the location of what is now the Bering Strait, while at the same time the corridor wasn't flooded because there was still a lot of water locked up in glacial ice. Thus, there was a window of time, perhaps a thousand years long, during which a migration from Asia would have been possible.

This simple picture has been modified somewhat by the discovery of sites occupied by humans before this window opened. The most striking and best verified of these is at a place called Monte Verde in southern Chile, but there are many more, including some in the eastern United States. As a result, our picture of the peopling of the Americas has become more complex, with an understanding that there were probably successive waves of immigration as people moved along the boundaries of the ice sheets in boats, probably from both Asia and Europe, in addition to the land migration from Siberia. Nevertheless, it is to this latter land migration that most Native Americans can trace their ancestry, because that was when the human influx started in earnest.

The first thing that the newcomers did when they arrived was to hunt many of the continent's large animal species to extinction. The mastodon is probably the best known of these animals, but fully three-quarters of the large animals on the continent disappeared, probably because of the skilled hunting of the newcomers. (In the words of my colleague Richard Monastersky, "Twelve thousand years ago Kansas looked like the Serengeti Plain on steroids.") This pattern of humans coming in and wiping out indigenous species is not confined to the Americas, but seems to be a universal behavior of our kind. When humans first entered Australia between fifty-three thousand and sixty thousand years ago, for example, every animal weighing over one hundred pounds became extinct.

Actually, there used to be a minor controversy among paleontologists and anthropologists about the connection between these extinctions and human hunting. You can think of it as a debate between "humans did it" and "climate did it." To my mind, however, the fact that we have evidence in two places—North America and Australia—and the fact that when the extinctions occurred, the climate was changing in different ways in the two regions—warming in America, cooling in Australia—pretty much decides the issue. So Native Americans, far from being different from the rest of us, behaved exactly as humans have always behaved. They exploited the most easily accessible resources of their environment (big game) relentlessly and, in this case, carelessly, because that was the easiest way to feed their children.

There is a parallel to this story in the modern fishing industry. Modern humans seem to be in danger of doing the same thing with the world's fish stocks. Because fish are "free," they are being taken from the oceans faster than they can reproduce. (In fact, fishing remains one area of human activity that has never gotten past the hunter-gatherer stage.) In the words of the Book of Ecclesiastes, "There is no new thing under the sun."

But Native Americas did much more than hunt big game. Over the millennia, they developed technologies that allowed them to manage their environments to the same ends (although not by the same means) as human beings did in Europe and Asia. Those ibis-haunted canals I saw in Peru are one example of this, but there are many others. It is, in fact, the sheer ubiquity of environmental management in the New World that has led anthropologists to begin to question the notion of the pristine wilderness.

Some anthropologists have argued that there have been two great centers of human civilization in the West—one centered in the Middle East and the other in the New World, in central Mexico. The dominant technological/industrial culture of the world today descends from the first, while the second was largely destroyed. The two cultures often took different approaches to problems like supplying themselves with food and managing their environments, and this has sometimes made it difficult for Western scientists to recognize the technologies developed in the New World.

Let's take North America, particularly the plains, as an example, and think about how Native Americans there solved the problem of supplying themselves with food. On the plains, the prime source of protein was buffalo and, to a lesser extent, other game. The European solution to this problem would have been to break the land up into individual plots and introduce domestic cattle as a source of meat. The Native Americans, on the other hand, used the technology they had—primarily fire—to modify the natural ecosystem so that it was a perfect grazing ground for buffalo, which they then hunted. In effect, they used fire to turn their environment into a game preserve for the purpose of raising meat.

The technique works like this: when the prairie burns, weeds, which have shallow roots, tend to be destroyed. The native grasses, on the other hand, have roots that go deeper, so they can sprout quickly once the fire has passed. This tender young grass is the preferred food of the buffalo, which then tend to congregate in areas that have been burned. By burning selected areas of prairie, people were able to get their protein to come to them.

You will recognize this process as a rangeland version of what I called protofarming in chapter 3. In a sense, though, it went beyond that, because it involved turning the entire natural ecosystem into a game preserve for the benefit of the human caretakers. Some scholars argue that the entire Midwest, from Illinois west to Nebraska, was a kind of natural artifact, an open prairie maintained by fire. It was only after the collapse of the Indian societies that oak forests—the ecosystem we think of as "natural"—invaded the region.

The evidence goes further than that. Archaeologists examining the early human settlement of the Isthmus of Panama have discovered an amazing fact. When they dig down to uncover sediments dating back to the first wave of humans, they find what is called a charcoal horizon, indicating widespread fires. "These guys just moved south," says Michael Moseley, distinguished professor of anthropology at the University of Florida, "burning as they came." Ecosystem management by widespread burning, in other words, seems to have been a feature of Native American civilization from the start.

Early European settlers witnessed some of this burning activity. The original Dutch settlers of New York (then New Amsterdam) used to take boats up the Hudson River in the fall to watch the fires, and the first settlers in Ohio found a landscape that looked like a park—they could actually drive their wagons between the trees in the open forest.

It has even been argued that the vast buffalo herds that figure so prominently in western mythology were what ecologists call an outbreak population. These populations, usually a symptom of a disrupted ecosystem, grow wildly when natural constraints are lifted. Buffalo herds, for example, grow at the astonishing rate of 30 percent per year if they are not culled. Without Native American caretakers to control their numbers, the argument goes, the buffalo population grew wildly, so by the time Europeans came on the scene, the buffalo had grown to the familiar "horizon to horizon" herds, later decimated by white hunters. This notion is supported by the fact that when Hernando de Soto led his army on a four-year trek through the American southeast in 1539, he never reported seeing a buffalo, while a century later, French voyageurs reported buffalo "grazing in herds on the great prairies which then bordered the river."

In any case, as far as North America is concerned, it is clear that what the Europeans called "wilderness" was actually a terrain managed by human beings to meet their own needs. And if this was true of the Northern Hemisphere, it was doubly true of the Southern, where Native American civilizations reached a level that equaled and, by some arguments, exceeded that of the Old World. There are three general categories of these civilizations that come to most people's minds when they think of South and Central America: the Aztec in the Valley of Mexico, the Maya on the Yucatán Peninsula, and the Inca in Peru. Each of these civilizations had precursors, successors, and contemporaries, of course, but we can think of these areas as three great centers of culture, somewhat analogous to Egypt, Greece, and Rome in Europe. The civilizations of western South America, which culminated with the short-lived Inca Empire in the fifteenth and sixteenth centuries, are a good example.

To get a better picture of this unfamiliar (to me, at least) terrain, I talked to Moseley, who knows the civilizations that grew up in the Andes and along the west coast of South America as well as anyone. He has the gruff, no-nonsense style of a man who has spent a good deal of his life in the field, far away from the niceties of polite academic society. He says what he thinks in no uncertain terms, a quality that endears him to authors looking for a juicy quote.

"In this part of the world," he says of western South America, "water has been the limiting factor since people arrived." Those canals I saw in the Peruvian desert are a lasting monument to that fact. The only water that reaches the west coast is carried by rivers fed by snow and rain high in the Andes. About five thousand years ago, people along the coast of Peru, who made their living by fishing, began practicing a simple kind of agriculture. They would flood fields with water, put in their plants, go back to fishing, and then come back to harvest whatever had made it through their absence. By this time, people in the region had domesticated gourds, chile peppers, and, perhaps, beans, so this sort of water management and farming could provide extra calories for people living in a difficult environment.

I felt a resonance when Moseley talked about this farming style, because those ancient Peruvian fishermen practiced what I used to call percentage gardening. I used to spend my summers in Montana, but before I left my garden in the Blue Ridge Mountains of Virginia I would do a thorough weeding, lay newspapers down around the plants, and cover everything with a thick coat of rotted sawdust from a local sawmill. I would then return two months later and harvest whatever had survived. I could usually count on a good late-summer tomato harvest, and maybe some water-melons. Gourds were always a good bet, too. Consequently, when Moseley told me about these early farming practices in Peru, I recognized them immediately.

This sort of protoagriculture provides a useful hedge against fluctuations in food supplies from hunting and gathering activities like fishing, but in the end, people who switched to full-time agriculture would eventually displace their kin who depended on fishing. The choice came for the early Peruvians in about 1800 B.C., which is the time when large-scale irrigation systems begin to show up in the floodplains of the west coast rivers. Over the next thousand years, this new technology spread, and we can document the switch to agriculture from fishing by observing the increase in dental caries in skeletons, caused by the increased amount of carbohydrates in the diet. Farming is a hard life, an unpleasant life, but at the same time it is the only life that gives a good chance of survival.

Irrigation systems spread from the narrow river valleys to the floodplains to the deserts as the population grew. By 1000 A.D., there was more land in Peru under irrigation than at any subsequent time until the late 1990s. Indeed, much of the modern Peruvian effort to reintroduce irrigation in its desert regions follows the old Indian canal system.

But those who choose to modify nature to ensure their survival have truly grasped a tiger by the tail. The price of survival is constant effort. Those rivers that brought the precious water down from the Andes also brought sediment that clogged the canals, while periodic El Niño events brought disastrous floods that destroyed them. But for a thousand years, the ancient Peruvians persisted, rebuilding their canals and going on. Finally, a three-century drought came, cutting rainfall in the Andes and destroying the delicate water balance on the plains. (Anthropologists speculate that this was the same drought that wiped out the Anasazi in the American Southwest, causing them to leave their pueblos.) At the end of the drought, in about 1400 A.D., the Incas expanded and founded their empire, which fell shortly thereafter to Spanish conquistadores. But in the river valley I saw, those old canals are still there, carrying out the function envisioned for them by engineers and builders long gone.

On the other side of the Andes, in wetter tropical climates, there is more evidence of human involvement in controlling the environment. Here, the problem for agriculturalists isn't the lack of water, but too much of a good thing. In western Bolivia, for example, rain and snowmelt from the Andes regularly turn areas the size of New York State into something resembling the Florida Everglades—a slow-moving sheet of water that eventually flows into rivers that carry it to the Amazon. Obviously, this sort of flooding would make farming difficult. The first settlers of this area, therefore, built up ridges of land, some hundreds of acres in size, to keep their fields above the encroaching flood waters—ridges that today support forest ecosystems very different from the surrounding grasslands. These ridges extend over an area of tens of thousands of square miles and shape the landscape they occupy. It is an example of what scholars are increasingly beginning to call "cultivated landscapes," a term that carries the connotation of human shaping and manipulation, rather than the operation of natural selection.

These kinds of ridges are found in millions of acres throughout the moist tropics. "They are the largest man-made structures ever constructed," says Moseley, but he hastens to warn that they were not all in operation at the same time. A visitor to the western Amazon a thousand years ago, in other words, would not have found all the ridges devoted to farming at the same time, even though over time all the ridges were used.

Farther east, scholars are beginning to argue that a different kind of protoagriculture was practiced. Basing their arguments on human detritus found in the excavation of caves that were occupied starting about eleven thousand years ago, they argue that humans actually treated the rain forest the way their North American brethren treated the prairie—they modified it for their own use. Scholars point out that conventional clear-field agriculture in a dense rain forest is extremely difficult if the only tools at your disposal are made of stone. On the other hand, the forests produce an astonishing variety of fruit and nuts. The idea is that the first human settlers began planting trees—in effect, turning the forest into an orchard, which they then tended. In 1989, William Balée of Tulane University estimated that 12 percent of the nonflooding Amazon rain forest is essentially the product of human effort. If anything, estimates are running higher these days. Indeed, Clark Erickson of the University of Pennsylvania argues that the Amazonian rain forests are the finest works of human art on the planet! As one author has pointed out, using the language of science fiction, the Indians were busy terraforming the Amazon Basin when the Europeans showed up.

Unfortunately, as sometimes happens in academic debates, the argument over cultivated landscapes seems to have gotten rather nasty. This can happen when a debate acquires an ideological dimension, especially if that dimension has to do with modern political correctness. I can remember, for example, the spirited debate that surrounded the suggestion (now widely accepted) that dinosaurs were driven to extinction by the impact of an asteroid. For some reason that I never understood, this got mixed up with the (then current) debate on nuclear winter and disarmament, and people who attacked the asteroid hypothesis were publicly accused of being in favor of nuclear war. (Don't ask me to explain this—I can't.) In much the same way, charges of colonialism and CIA employment get leveled at people who argue for the existence of cultivated landscapes. Once again, I don't get it, but there it is.

Ideology aside, what is emerging from the study of early human occupation of the New World is a very different picture from the one painted in schoolbooks. Native Americans were not a meek people, happily accepting what nature had to offer. Like their fellow humans in other parts of the world, they actively intervened in their environment, changing it to suit their purposes. In the process, they created a landscape in the Americas that was about as far from "pristine" as it could be. From the Great Plains in the north to the rain forest in the south, the stamp of human intervention and ingenuity was on the land when Europeans first arrived on these shores.

And when Europeans showed up, there were a lot of people in the Western Hemisphere—by some estimates, more than were in Europe. Exactly how many Native Americans there were also remains a subject of spirited academic debate, with estimates ranging from 1.8 to 18 million, but those numbers are certainly higher than the conventional wisdom of just a few years ago. And as they modified their environment, Native Americans expanded their population to the limit of their technology, as humans always do.

And life in the New World was good. The Aztec capital of Tenochtitlán, for example, had a larger population than Paris. When John Smith first visited Massachusetts in 1614, he wrote that the land was "so planted with Gardens and Corne fields, and so well inhabited . . . (that) I would rather live here than any where." In a recent straw poll of seven leading anthropologists, author Charles Mann found that all seven would, if given the choice, have chosen to live in the Americas in 1491, rather than in Europe.

Mythos Versus Science

If this is so, then how did the myths of the noble savage and the pristine wilderness arise? How could European explorers not have noticed the advanced cultures and the people that were all around them?

In a word, disease.

Let me begin this discussion by pointing out that in the sixteenth century, no one had any idea where disease and plagues came from. Louis Pasteur and the germ theory were still three centuries in the future, and plagues were often regarded as punishment from God rather than as natural phenomena. Nevertheless, diseases played a crucial role in the post-contact history of the New World.

The great land mass of Eurasia and Africa contained not only the birthplace of humanity, but the largest collection of living humans in existence. Diseases that developed in one area often worked their way around all three continents (think of the Black Death). Over the centuries, then, European populations were exposed to many diseases. The workings of natural selection tended to weed out those whose immune systems couldn't counter those diseases, so over time Europeans developed full or partial immunity to many viruses and bacteria.

Native Americans, on the other hand, were relatively sheltered and isolated from the great continental germ pool across the Atlantic. They had not been exposed to many of the common diseases with which the ancestors of the European newcomers had dealt for centuries. Consequently, when those diseases were introduced into the native population, there was almost no resistance to them. The result: the greatest epidemiological disaster of all time.

Starting with smallpox, the list of diseases that decimated the native population reads like a medical textbook—typhoid, measles, mumps, bubonic plague, whooping cough, followed later by cholera, malaria, and scarlet fever. One after the other, they took their toll. Given what we now know about bacteria, we realize that it wouldn't have taken much to get a major epidemic started. A single infected European could start a plague that wiped out millions—indeed, we think that after smallpox was brought to Mexico by a single person in 1525, it spread south until it had killed half the population in the Inca empire. Hernando de Soto's expedition may have done the same sort of damage in another way. Among their provisions when they landed were about three hundred pigs, some of whom undoubtedly escaped and spread diseases to wildlife in the surrounding forests, from which diseases were transmitted to native populations. In any case, while de Soto's men reported Arkansas to be "very well peopled with large towns," when René-Robert Cavelier, Sieur de La Salle, came through in 1682, he didn't find a single village in two hundred miles of travel. Clearly, something horrendous had happened in the 150 years that intervened between these two reports.

What had happened, of course, was a catastrophic collapse of the native population. Estimates of the depth of the loss vary, from upper estimates of a 95 percent casualty rate to lower estimates of 60 percent (a number derived from modern data on populations exposed to new pathogens). Frankly, although these numbers are hotly debated, which one you choose makes little practical difference. Imagine what would happen to your hometown if two-thirds of the people in it suddenly vanished. Who would drive the buses, run the power plants, staff the food stores? When I go through this exercise for my native Chicago, I see a city in collapse, with survivors scrambling to find enough food and shelter to stay alive. It's not even too hard to imagine a return to subsistence farming in such a case.

I suspect the impact of new diseases on the native populations of the Americas was similar to this. Once-great cities crumbled when there was no one there to care for their infrastructure. Armies decimated by disease faced unfamiliar mounted foes in armor and thus lost empires. And, perhaps worst of all, no one—including the Europeans—had any idea of what was happening.

The noble savage and pristine wilderness myths, then, arose because Europeans first became aware of Native Americans during an anomalous time in history. *Of course,* it looked as if the Indians lived in small groups—their population had been decimated by disease. *Of course,* they lived lightly on the land—their technological infrastructure had collapsed along with their population. *Of course,* they had little impact on their environment—

the plagues had destroyed their ability to maintain their large-scale ecosystem management projects.

In the end, the myths arose because Europeans took a "snapshot" of Indian culture during a period of extreme stress. In point of fact, the "wilderness" that the Europeans found in the New World was nothing less than a continent that was in the process of being shaped by human beings to meet human needs. There is no preferred state of nature to be found here, no lost ideal to which we can all return. There is nothing but ourselves seen in the mirror of another culture.

And where does the pristine wilderness fit into all of this? In the pithy words of Mike Moseley: "Pristine wilderness? Give me a goddamned break!"

7

The Myth That We Are
Poisoning the Earth

My introduction to the subject of this chapter came a number of years ago, when I was having lunch with my nephew, who was then in seventh grade.

"We're poisoning the planet," he informed the family gathered around the table.

Somewhat taken aback, I asked him precisely how we were accomplishing this task. He looked at me uncomprehendingly.

"We're poisoning the planet," he said scornfully, "everybody knows that, Uncle Jim."

Since then, I have learned that the poisoning of the planet is indeed one of those things that "everybody knows," and, like many things that "everybody knows," it requires some careful thought to figure out exactly what it means and exactly how true it is. My friend and colleague Berkeley biochemist Bruce Ames refers to this part of pop ecology as the Rachel Carson myth, after the author of the groundbreaking 1962 book *Silent Spring*. I will define the myth in this way: until the start of the industrial revolution, the environment was free of toxic and carcinogenic chemicals. Since that time, however, human beings have been flooding the biosphere with all sorts of noxious substances against which humans and other life-forms are defenseless. As a result, the biosphere has been devastated (*Silent Spring* was mainly a diatribe against the effects of DDT on songbirds), and humans have suffered incalculable adverse effects to their health.

Unlike the myths of stability and the pristine wilderness, which have virtually no basis in fact, this particular myth is partly true and partly false. There is no doubt that human activities have introduced harmful substances into the environment. The real question before us is twofold: precisely, and to what extent, do specific substances introduced by humans into the environment cause damage to human health and to other living things, and what should be done in each case?

This is an important question, because how we regard the question of chemical pollution will have a powerful effect on our management goals for the planet. If we believe that we are, indeed, being flooded by substances that pose a deadly threat to both us and the environment, then issues of cost effectiveness do not enter the picture. The threat would have to be removed whatever the cost. If, as I shall argue, the picture is more subtly nuanced than that, then the best management technique would involve the regulation of threats seen as serious, with less attention paid to lesser ones. In particular, it would involve a careful assessment of the risk posed by a specific substance, followed by a decision as to whether that risk is sufficiently high to warrant regulation. If it is, then another analysis has to be done to determine just what sort of regulation, from outright banning to specific controls, is appropriate for that specific substance. This isn't the stuff of bumper stickers and protest rallies, but it's where we need to go.

In what follows, I will be using some notions that might be unfamiliar to some readers, so I want to give fair warning here. First, I will be taking a point of view suggested by modern biology, in which we look at the effects of pollution at the molecular level, asking just how a particular substance affects the delicate chemical dance that constitutes what we call life. This is an important departure, because by concentrating on what is happening at the molecular level, we can get around the (ultimately unimportant) distinction between "natural" and "artificial" substances.

The second point has to do with the troublesome concept of risk. We all want to eliminate risk from our lives, but the world being what it is, we cannot. There is, for example, a (very tiny) risk that in the next minute a meteorite will enter the atmosphere and kill you. Risk is simply an unavoidable fact of life. Indeed, all responsible participants in environmental debates agree that risk can never be entirely eliminated from human life.

The question is how much extra risk we add to our lives by allowing certain substances to enter the biosphere, and whether that risk is justified by the benefits we derive from that substance. This is often a complicated subject, since those who benefit and those who assume the risk may not be the same people. Nevertheless, we can often make some sense by comparing risks—we might argue, for example, that being exposed to a certain amount of substance X introduces a risk equivalent to driving one hundred miles without a seat belt. This sort of comparison helps us to visualize small risks and is, in fact, a staple of the environmental debate.

It is important to realize that while scientists can (and do) produce quantitative estimates of the amount of risk encountered in specific situations, the question of how much risk is acceptable is highly personal, and its answer is definitely *not* a scientific one. Some people (like me) tend to be cautious and risk-averse; others tend to be risk takers. People in these two groups will often agree on the amount of risk involved in a situation while disagreeing on whether that amount of risk is acceptable. This is simply another fact of life.

Finally, the recognition that life cannot be risk free leads us to consider another two-part question: given that artificial chemicals can interfere with the chemistry of life, what natural chemicals have the same effects, and how does the added risk imposed by manufactured substances compare to the risk already imposed by nature? The idea that nature in and of itself can be malevolent is largely ignored in public discussions of environmental issues, but becomes very important when you discuss risk.

Unfortunately, this aspect of risk analysis is largely ignored in public debates (although not in debates among scientists). In the full-blown "litany" of pop ecology, every manufactured chemical poses a risk that is essentially infinite, while natural chemicals pose no risk at all. We have to do better than that. In our future as planetary managers, we have to be able to make quantitative comparisons of different kinds of risk (something scientists already know how to do) and use them to drive policy decisions. This will not be glamorous work—no movie star is going to appear in front of a congressional committee to argue that someone's numbers are off by 30 percent—but it is the only way to get to reasonable solutions to our problems.

As I did in the opening chapter of this book, I will look at three cases to talk about issues involved with chemical pollution. First, I will talk about pollution by heavy metals and persistent organic chemicals, where the risk to humans is reasonably well documented and where regulatory systems are already being put in place. Then I will look at the problem of the cancer threat associated with agricultural pesticides where, in my opinion, the threat to humans has been vastly overblown. I will also use this subject to show, in one small example, what a thorough analysis of an environmental problem looks like and, scientists being what they are, how disputes over details can affect ideas about policy options. Finally, I will look at the history of the use of DDT, which started life as a hero and wound up as a villain, to set the stage for the moral and ethical arguments that will be taken up later in the book.

At bottom, all living things are essentially chemical factories in which a delicate interplay goes on as molecules come together and are broken apart. Furthermore, how these molecules carry out their function depends primarily on whether or not they fit together—on their geometrical shape.

Introducing any substance into this dance will change the way that the molecules work. Sometimes this can have a beneficial effect, as when a medicine supplies a molecule that the body isn't making or an antibiotic removes a harmful organism. Other times it has a negative effect, as when the new substance attaches to a particular molecule and blocks some important function. In this case, the new molecule may cause sickness or even death. But harmful or beneficial, the new molecule does what it does because its shape allows it to alter the interactions between molecules in your body.

Let me stress this point. What matters in the molecular world is the shape that a molecule has, not where it came from. In particular, once we are into the molecular maelstrom, the question of where the molecule came from becomes completely irrelevant. To put this bluntly,

In the molecular world, the distinction between "natural" and "artificial" is completely irrelevant. All that matters is geometrical shape.

To take just one example among many, consider the action of a "natural" substance like curare. It affects the action of molecules called neurotransmitters, which pass signals from nerve cells to muscles. Even in small doses, it causes paralysis and death. This is very similar to what a deadly nerve gas like sarin does. They are both deadly, yet one comes from plants in the tropical rain forest while the other is made in laboratories. One is "natural" while the other is "unnatural." Yet at the molecular level, they both produce deadly results by affecting the action of the same neurotransmitter. (While both curare and sarin affect the operation of the neurotransmitter acetylcholine, there is a difference between the way the two act. Curare blocks the reception of the neurotransmitter in neurons, while sarin prevents the molecule from being broken down, so the nerve signal is sent repeatedly, a process that leads to muscle failure.)

When we talk about dangerous substances in the environment, then, we have to keep in mind that those substances can come from either the process of natural selection or manufacture by humans.

Carcinogens and DNA Repair

In general, we can classify the most important dangerous molecules in our environment into two groups—those that cause immediate and direct harm (toxins) and those that cause long-term harm by altering DNA and causing cancer (carcinogens). Toxins, in general, cause harm relatively quickly, so their effects are easy to see. Carcinogens, on the other hand, are more difficult to detect, since cancers may take years or even decades to develop after exposure. Nevertheless, we have accumulated a great deal of information about both kinds of molecules in the environment.

One of the earliest cases of an environmental toxin coming to public awareness involved the metal mercury. Normally found in the form of a silvery liquid, mercury has a number of industrial uses. It has long been known that people exposed to mercury can be adversely affected—the phrase "mad as a hatter" arose because mercuric nitrate was used in making felt for hats, and when it was absorbed through the skin over long periods of time, it produced marked changes in behavior.

In the early 1950s, people, birds, and household cats in the town of Minamata in Japan were afflicted with a wasting disease that resulted in paralysis and, in some cases, death. The disease was traced to mercury emissions from a nearby factory that, through the action of bacteria in ocean sediments, had been incorporated into organic compounds. They entered the food chain and were eventually concentrated in fish. Fisher-men, birds, and cats, all of whom ate a great deal of fish, also ingested a great deal of mercury compounds, which caused lesions in their central nervous systems and produced what is now called Minamata disease. Since this discovery, the discharge of toxic heavy metals is now rigidly controlled in the industrialized world.

Another class of substances that are monitored and restricted are the so-called persistent organic pollutants (POPs). These are molecules that remain in the environment for a long time and hence can be carried across national boundaries by wind and currents. They include classes of mate-rials like polycyclic aromatic hydrocarbons (PAHs), dioxins, and polychlo-rinated biphenyls (PCBs). There is a fair amount of data, some better than others, that suggest that various members of this class of materials can act as carcinogens or cause damage to the immune and reproductive systems of humans who suffer high levels of exposure. Because of this, in May 2001, the international community signed an agreement called the Stockholm Convention on Persistent Organic Pollutants, and these materials are now to be monitored and restricted (and, in some cases, phased out) over the coming years.

The Stockholm Convention, in many ways, represents one way in which I believe that the process of planetary management will be carried out. A problem is recognized, enough scientific knowledge is accumulated to identify the threat and the cost of mitigating it, and the international com-munity decides that the costs and benefits are such that steps toward miti-gation should be taken. In this, it is similar to the Montreal Protocol of 1987 (expanded and updated in the 1990s), which ultimately banned chemicals called chlorofluorocarbons, which had been shown to be causing damage to the ozone layer. These treaties illustrate the way in which chem-ical pollution will be managed in the future.

There are also many documented cases in which materials introduced into the biosphere have harmed other species, but not humans. Agricul-tural fertilizers washed into ponds in California, for example, can be absorbed through the skin of amphibians like frogs, decimating local pop-ulations. The use of DDT in the postwar era caused damage to bird popu-lations—a subject to which we will return below. But in terms of public awareness and public fears, it is surely the introduction of carcinogens that plays the most important role in the discussion of chemical pollution.

Carcinogens work by inducing changes in the DNA code in cells. If the damage to DNA occurs in the genes that control the delicate regulation of cell growth and death, the cell may begin to divide without the normal checks and balances that govern body architecture. The result: a tumor that, if left untreated, will destroy its host. Normally, it takes a number of "hits" to a cell's DNA before a cancer develops, and in some cases we can say exactly what sequence of damage is needed to induce a tumor. In the case of colon cancer, for example, it requires hits to genes on four separate chromosomes (five, twelve, seventeen, and eighteen) to turn a benign polyp into a tumor.

We call something a carcinogen if it can cause these sorts of changes in DNA. Chemicals are one type of carcinogen, but there are others. Radiation and even heat can cause damage to DNA. Surprisingly, the greatest number of carcinogens facing human cells do not come from outside the body, but are the normal by-products of human metabolism.

In order to generate the energy it needs to run its chemical operations, each cell in your body has to metabolize molecules in the food you eat. Think of this process as the cell "burning" sugars the way a power station burns coal. And just as coal plants are normally located far from population centers to minimize the damage from the by-products of burning, energy production of cells is isolated in specific structures called mitochondria. Despite this fact, potentially destructive molecules from the burning process can (and do) damage the cell's DNA. There is no way to prevent this—if you want the energy, you have to accept the damage.

Consequently, the cell has many mechanisms to repair damage to its DNA. Specialized molecules constantly patrol up and down the double helix, and when an abnormal bulge (signaling damage) is detected, they call in other molecules that initiate repairs. In essence, these molecules snip out the side of the DNA "ladder" that has been damaged and then, using the standard machinery of the cell, rebuild it correctly.

To find out more about DNA repair, I talked to Bruce Ames, professor of biochemistry at the University of California at Berkeley. A thin man with wavy gray hair and a neatly trimmed mustache, Ames is a real scientist's scientist. Early in his career he developed what has come to be known as the Ames test—a quick, cheap way of determining whether or not a particular chemical causes damage to DNA in bacteria. More recently, he has become interested in the issues surrounding DNA repair and the presence of carcinogens—both natural and man-made—in the environment.

He studies DNA repair by looking for the pieces of excised DNA when they are flushed out of the body in the urine. "Everybody in Berkeley is high on something," he is fond of saying, "and I'm high on urine." His work has uncovered a remarkable picture of the functioning of DNA in human cells. "Every cell in your body takes about ten thousand hits to its DNA every day," he says. And, as I pointed out above, most of those hits are caused by chemical reactions involving the by-products of the cell's own metabolism.

This is an amazing result that speaks directly to an important implicit assumption in pop ecology—the assumption that DNA is somehow a fragile, vulnerable target for environmental carcinogens. Actually, the real picture of DNA in cells is much more dynamic. It is a picture of a molecule constantly under assault, while at the same time constantly repairing itself. The life of DNA is a contest between the mechanisms of damage and the mechanisms of repair. Molecules from the by-products of metabolism and from the environment are constantly stressing the molecule, while molecules in the cell are constantly repairing the damage.

In fact, there seems to be a hierarchy to the way DNA is repaired. We know that only about 5 percent of human DNA is actually taken up by genes, and only a fraction of those genes are active in any cell. Thus, not all damage to DNA is equally dangerous to the organism. Just as you would repair the roof on a storm-damaged house before you replaced a carpet, the cell repairs active genes first, since they are the most critical part of the DNA. Other repairs get a lower priority.

If you think about the process of natural selection described in chapter 2, this picture of DNA makes sense. Living systems had to evolve in a dangerous environment, one that was full of potential carcinogens. Those cells that could repair damage to their DNA would obviously be more likely to survive long enough to pass on their genes to the next generation. Thus, over time, you would expect that complex repair mechanisms would develop, since each improvement would grant an organism a definite edge in the game of natural selection.

But that, of course, raises another question. We have seen that some carcinogens are a natural by-product of metabolism, but are there others in the environment? Is it possible, in other words, that humans have been subjected to attack by environmental carcinogens since we first appeared on the scene millions of years ago? Are there carcinogens in the environment that have nothing to do with industry or even with human activity? We know that the answer to these questions is yes.

Consider natural selection from the point of view of a plant. It has vegetable competitors—other plants that try to co-opt its sunlight and nutrients. It competes with these by the standard methods of natural selection—by growing taller, by using sunlight more efficiently, and so on. But it also has predators, mainly insects, that want to use it for food.

How can it deal with these? A plant can't run away, and it has no teeth or fangs to fight back (although it may have thorns). The only weapons at its disposal are chemical. If it can manufacture molecules that will harm insects that try to eat it, it can increase the odds of surviving. And this is precisely what plants have done. "Plants are better chemists than Dow and Monsanto," says Bruce Ames, referring to the wide range of chemicals plants produce to protect themselves against insect predators.

Of course, a chemical that kills insects is a pesticide, so what we are saying is that every plant, just to survive, has had to evolve a series of natural pesticides. Humans, over the past century, have added to this arsenal with their own chemicals aimed at killing insects—what we can call manufactured pesticides. The question of "poisoning," then, comes down to the question of whether the manufactured pesticides are more harmful to us or the environment than the ones that occur naturally.

"About half of man-made pesticides have been shown to be carcinogens," says Ames, "and about half of the natural pesticides are the same." His data come from the standard experiments in which laboratory mice are exposed to chemicals and then monitored for incidence of cancer. This means that once pesticide molecules get inside the body, about half of them have the ability to cause the changes in DNA that can lead to cancer, regardless of whether the original molecule came from nature or from a factory. The question about the danger of cancer from pesticides comes down to one of quantity—when you eat an apple or a carrot, how much pesticide from natural sources are you ingesting compared to the pesticides that were used by farmers while the plant was growing?

"Roughly speaking," says Ames, "99.99 percent of the pesticides you take in each day come from natural sources." Eating an organic apple, in other words, delivers almost as much in the way of carcinogenic molecules to your cells as eating an apple that has been grown conventionally. "When you eat cabbage," Ames continues, "you ingest forty-nine different natural pesticides."

To make his point, Ames produces a list of fruits and vegetables, along with the known natural carcinogens found in each. Anything you've ever eaten is on the list—even the parsley, sage, rosemary, and thyme celebrated in the old song. Reading the list of natural pesticides reads like a nightmare warning on a food label:

Acetaldehyde
Benzofuran
Benzo(a)pyrene
1,2,5,6- Dibenzanthracene
Hydroquinone
Toluene

And that's just from a cup of organically grown coffee!

Ames makes another point about pesticides. "The best defense against cancer," he says, "is eating fresh fruits and vegetables." The reason is that these foods contain molecules known as antioxidants. They block the action of the most common cause of damage to DNA—the effects of the by-products of metabolism. "Anything that pushes up the price of fruits and vegetables will have the effect of reducing the amount that poor people can consume, and will therefore actually increase the incidence of cancer in a population," Ames continues. Since the whole point of using pesticides in agriculture is to produce higher crop yields (and therefore lower prices), it follows from this argument that the periodic scares about pesticide residues to which the public is subjected by well-meaning environmental advocates may actually cause a great deal of harm to people.

When I make these arguments to friends and family (particularly to my daughters, who are, respectively, medical and law students), the most common reaction is to say something along the lines of "Even if what you say is true, why should I accept *any* additional risk at all?" The answer lies in the sort of risk comparison discussed at the start of the chapter. If we accept Ames's numbers, the amount of manufactured pesticides the average person would receive from conventionally grown fruits and vegetables each day is about the same as the amount of natural pesticides he or she would get from taking one bite of an organically grown carrot. Since most people would accept the latter risk without a second thought, I conclude that even for a cautious individual like myself, the added risk of cancer from conventionally grown foods is negligible. Couple this with the benefit of having cheap fruits and vegetables available to all people, and the choice (for me, at least) is obvious—the benefits of agricultural pesticides far outweigh the risks, at least as far as carcinogenity is concerned.

But as you might expect, there is far from unanimous agreement on all aspects of this argument among scientists. The existence of DNA-repair mechanisms is about as well documented as something can be in science—the prestigious journal *Science* gave these mechanisms its coveted "Molecule of the Year" award a few years ago. The existence of natural carcinogens is also well known, although before Ames's work, I don't think their full extent was appreciated.

The debate on pesticides centers on two issues: (1) are there effects other than those associated with carcinogens that we should be worried about, and (2) are the numbers used in the argument really solid?

We can, for example, ask about the toxic effect of agricultural chemicals. What about the fact that many of them are actually neurotoxins that attack the insect nervous system? Given the similarities in chemistry among living things, couldn't those chemicals have the same effect on humans?

"No one has ever looked for natural neurotoxins in plants," says Ames, "but given the logic of the situation, I'm sure that they're there. If humans can develop them, plants can, too."

The term "logic," of course, refers to the rules of natural selection. If pesticides containing neurotoxins are really effective, as they appear to be, then any plant that can manufacture those particular molecules will enjoy an advantage in the game of natural selection. You would expect plants with the genes for that particular type of molecule to prosper, so the human diet would be as full of that particular pesticide as it is of carcinogens. We know that animals (including humans) that eat plants have developed defense mechanisms to repair damage to DNA caused by natural pesticides, and we should expect that if the plants also produce neurotoxins, analogous defense mechanisms should have developed for them as well. The fact that we don't know what those mechanisms are tells us more about the way that scientific research is managed than it does about the mechanisms themselves. It is simply not a high priority in today's scientific world to investigate questions like this, so progress in this field has come slowly. If we are to have rational management of chemical risks, however, this sort of work will have to be done. At the moment, however, the comparison of natural and man-made pesticides and toxins remains an unsolved problem.

As far as the debate about numbers is concerned, as with many other issues discussed in this book, it involves messy details. The main issue is the calculation of the natural versus man-made pesticide load. How firm is the 99.99 percent number quoted above? If this number were significantly smaller, then agricultural pesticides could be adding significantly to our overall cancer risk.

This debate quickly descends into questions like how much of what sorts of natural pesticides are included in the average diet—how much basil does the average person consume versus how many mushrooms, for example. When I look at the numbers, though, it's hard to see how Ames's basic conclusion could be changed significantly. Even if his numbers overestimate the amount of natural carcinogens by a factor of two or three, the natural load of carcinogens still swamps the effect of manufactured substances. The difference between the added risk of manufactured pesticides being equivalent to one bite of organic carrot per day (as Ames claims) or being equivalent to an entire organic carrot per day (which is my interpretation of what his critics claim) doesn't change the basic structure of the argument. Either way, we are at far more risk from the carrot than from the manufactured pesticides.

The discussion of DNA repair and natural carcinogens is meant to show that the real world is a lot more complicated than that portrayed in pop ecology. In at least one important case where we have reliable information, the case of agricultural pesticides, it appears that the things that "everybody knows" about the environment and the food supply may not be exactly true.

The Story of DDT

Malaria is one of the great scourges of humanity. Today it produces over three hundred million cases of acute illness and over a million deaths each year. The disease is caused by a genus of microbes known as *Plasmodium* and is transmitted to humans through the bites of mosquitoes, which pick up the parasite from one person and pass it to another. Although there are about twenty-five hundred different species of mosquitoes in the world, only mosquitoes of the genus *Anopheles* (about sixty species) can actually transmit the parasite to humans. One way to combat the disease, then, is to get rid of the mosquitoes.

All this is by way of introduction to some rather startling events that took place in a laboratory in Switzerland in the late 1930s. A chemist by the name of Paul Müller was looking for substances that would, in essence, make better mothballs. He would coat a glass tube with a candidate substance, fill it with houseflies, and see if they died. If they did, the substance was investigated further. If not, it was dropped from the search process. One day he tried a substance called dichloro-diphenyl-trichloroethylene—DDT, for short—and found that it had a truly amazing ability to kill insects. Minute traces of the substance would kill essentially every insect that came into contact with it. (We now know that DDT works by affecting the nervous system of insects, causing the nerve cells to continue firing until the insect goes into a spasm and dies.)

The company for which Müller worked quickly sent samples to its New York office, from whence it found its way to the U.S. Department of Agriculture. By then it was 1942, and the country was embroiled in a war. The development of an effective insecticide was suddenly a high priority by the military, because insect-borne diseases were rendering American soldiers in the Pacific incapable of fighting. In that year, for example, the First Marine Division was pulled out of combat and sent to Australia to recuperate because ten thousand men (of a total force of seventeen thousand) had contracted malaria. Typhus (transmitted by lice) also affected the troops.

Scientists quickly tested the DDT on rabbits and mice to see if was toxic, and then on human volunteers. As far as they could see, it harmed only insects. To military planners, DDT was little short of a miracle—cheap, easy to use, effective. Commanders in the Pacific Theater routinely called in DDT sprays the way they called in air strikes against enemy positions.

The impact of the chemical on the postwar world was equally profound. It's always difficult for people today to understand the way that long-conquered and almost forgotten diseases dominated earlier times. As an opera buff, I am constantly reminded that tuberculosis was a disease as deadly as AIDS in the nineteenth and early twentieth centuries—if a character in an opera coughs in the first act, you know she will die in the last. I can still remember being forbidden to leave our yard during childhood polio epidemics in Chicago. It is also hard to realize that malaria was once endemic in Europe, Asia, the Caribbean, and the southern part of the United States. There is one reason that it is no longer found in many of those places—DDT.

After the war, a massive antimalarial campaign was launched. The strategy was ingenious. The way that the disease is transmitted is simple: a mosquito drains blood from someone who has the parasite. The mosquito then seeks out the nearest vertical surface and stays there for several hours, draining its gut. Only then will if fly off to bite someone else, passing on the parasite in the process. If that vertical surface (a bedroom wall, for example) is coated with DDT, the mosquito will absorb enough during its resting period to kill it. The result: the infection cycle is broken and the disease does not spread.

Since people typically rid their bodies of the malaria parasite in a period of three years, if you can keep up the spraying campaign for this long, you can eliminate the parasite from an area. If the mosquitoes come back after that it doesn't matter—they may bite, but there are no more microbes to pass around. The disease will have been eradicated.

Following precisely this kind of procedure, malaria was, in fact, eliminated from Europe and the United States, as well as from most of the Caribbean and large swaths of Asia. In India, where an estimated seventy-five million people contracted the disease (and eight hundred thousand died) every year, the fatalities dropped to zero by the early 1960s. Estimates of the number of human lives saved by this single chemical run to the tens of millions. DDT clearly saved more lives than any drug or chemical in history. In 1948, Paul Müller won the Nobel Prize in chemistry for his discovery.

At the same time, farmers, particularly in the western United States, began to use the new "miracle" insecticide on their fields, spraying it indiscriminately. After all, it worked beautifully, and no one was thinking about what else DDT might be doing once it got into the environment.

Then came *Silent Spring*. Suddenly, all the qualities that had made DDT so effective in the battle against malaria came back to haunt it. Its main advantage was that it was persistent. This meant that it would retain its potency for a long time and continue to kill mosquitoes months after it had been applied in a house. It also meant, however, that once it entered the environment it had time to work its way up the food web, from plants to animals like birds. For example, studies in Long Island Sound showed that zooplankton had eight hundred times the concentration of DDT found in the environment and that by the time the DDT had worked its way up through various fish to seagulls, the magnification of DDT was over two hundred thousand. Carson quotes, with approval, a letter from a woman in Alabama who asks, "What is man doing to our perfect and beautiful world?"

To be fair, there was a huge difference in the way DDT was used by public health people and the way it was used in agriculture. Spraying the walls of a house every six months requires much less of the chemical than spraying a field. It has been estimated, for example, that the amount of DDT needed to treat all of the houses in a country like Guyana is less than the amount that would typically be used on a large cotton farm. And even though many of the claims made in *Silent Spring* were overblown, the central message—that we were using DDT without thinking about its environmental consequences—remains valid.

We have to note, however, that whenever a major engineering project is undertaken, there will be mistakes at the beginning. Think of the early days of the Apollo program, which eventually put human beings on the surface of the moon. In the beginning, one rocket after another blew up on the launchpad, until finally the engineers got things right. I think of the unintended consequences of the use of DDT as analogous to those early explosions. Just as those explosions taught us how to build our rockets, the experience with DDT taught us something about how to carry on the chemical battle against insects. Today, pesticides are engineered to break down once they are sprayed, and they tend to be targeted much more closely to specific pests and crops. Part of the Stockholm Convention on Persistent Organic Pollutants includes restrictions on the use of DDT and a request that the few countries that still manufacture it find ways to phase it out.

Every decision has consequences. We decided to use DDT, and that had consequences. Had we decided not to use it, however, there would have been consequences as well. How "perfect and beautiful" would the world be with ten million people dead from malaria, for example?

Having said this, and as an exercise leading up to the discussion of how the planet will be managed later on, let me pose a hypothetical situation to you. Suppose that when DDT was discovered, we knew exactly what the costs and benefits of its use would be. Suppose someone told you that if we used it we would save ten million human lives and prevent untold human suffering, but that at the same time we would defile our "perfect and beautiful world" by eliminating songbirds, so the spring would indeed be silent. Suppose we were standing outside of a hut in a malaria-infested region of Africa and I handed you a sprayer full of DDT solution. If I left and came back in half an hour, would the sprayer be full or empty?

While you are pondering that question, let's consider what, in the end, we are to make of the notion that we are poisoning the planet? It is clear that some of the substances that human beings release into the environment—heavy metals and persistent organic chemicals, for example—have the potential of causing harm both to us and to the rest of the planet. They do, indeed, constitute a kind of poisoning, one that will have to be managed carefully in the future. As we have seen, there already exists an international political framework to carry out this task for some substances, a framework that will probably serve as a model for future managers of our planet.

On the other hand, it appears to me that the claims about the cancer threat associated with the most widely publicized and best-known "poisoning"—the use of agricultural pesticides—really don't stand up under close scrutiny. The fact is that, in this area, human activity closely mimics the strategies that plants have developed to protect themselves from predators, so the cancer scares associated with pesticide use have little substance to them. Human DNA is not a fragile, delicate thing. Like everything else in the natural world, it is tough, resilient, and, to a large extent, capable of taking care of itself.

In the end, then, we cannot make blanket statements about the introduction of chemical substances into the environment, but must examine each in terms of the risks it imposes, compare that to the risks that occur naturally, and come to a decision. This process doesn't require much new science, but does require the political will to accept and deal with a complicated world.

8

The Question of Extinction

Sometimes a statement is made so often that it assumes the status of received wisdom, even though it may not be true. I will argue that the claim that human actions are causing the most dramatic extinction of living things since the events that led to the demise of the dinosaurs sixty-five million years ago is in this category.

The validity of this claim is an important question for a number of reasons. For one thing, it touches on a deep problem in current debates on the environment—a problem related to the Fallacy of the Snapshot I described in chapter 5. Large numbers of scientists began to study extinction and the planet's biodiversity during the late twentieth century. We may be able to measure or estimate extinction rates today, but we can have no idea if those rates are unusual unless we can compare them to extinction rates in the past. It's not enough, in other words, to say that a certain percentage of species on the planet are going extinct—you have to know whether this rate is unusually high or not, and to decide that, you have to know what the rate has been in the past.

It's a little like looking at the weather. Several days of continuous rain would not be at all unusual in places like Seattle and London, but it would be highly unusual in Tucson or Phoenix. What's unusual depends on past history.

Furthermore, arguments about extinction rates have important implications for environmental policy and for the problem of managing the planet. If current extinction rates aren't much out of the ordinary, then there really isn't much that has to be done. If they are high, however, the debate shifts to what measures need to be taken to deal with a crisis. Among the things that would have to be on the table are questions such as which species are to be protected and which are to be allowed to vanish, which ecosystems are to be preserved, and what principles should guide our preservation efforts.

Finally, on a more philosophical note, statements about massive extinctions play into the kind of self-hating view of human beings that one encounters occasionally in environmental advocates—a view that leads to rhetoric in which humans are compared to a disease on the planet Earth, with the not-so-hidden message that the world would be better off without us. With respect, I beg to differ from this point of view, and I want to think very carefully about any argument that strengthens it.

Having said this, I have to add that with this discussion we are getting into a new area in our exploration of the world we actually live in—an area where the scientific evidence for or against any argument is ambiguous or even nonexistent. Scientists for whom I have enormous personal respect can be found on both sides of the extinction debate, so I would prefer to label it an "unanswered question" rather than a "myth."

One problem is that scientists from different fields tend to approach this question differently. Just as a lawyer, an economist, and a political scientist might approach a question like unemployment policy from different viewpoints, so, too, do ecologists (who study current ecosystems) and paleontologists (who study the record of the past) differ in the way they approach the issue of extinction. Ecologists tend to base their arguments on things like habitat destruction and theories about the way that organisms respond to it, while paleontologists tend to look at the fossil record.

In this case, I think that the paleontologists have it right. To deal with the question of whether current extinction rates are unprecedented, we have to ask, "What is normal—what does nature herself do without human intervention?" To do that, we have to go back to the subject of the second chapter and talk about the role of extinction in the evolution of life.

Extinction in the Fossil Record

"Extinction is forever," says the T-shirt, and, like many statements on T-shirts, this one is true—sort of. When you look at the fossil record—the annals of past life on the planet—you find that virtually every plant or animal that once had its day in the sun is now gone. Paleontologists estimate that 999 out of every 1000 species that ever lived are extinct. Some of these species gave rise to new species that succeeded them, as the australopithecines that walked the savannahs of Africa three million years ago eventually gave rise to modern *Homo sapiens.* Others simply disappeared, like many of the strange beasts whose fossils survive in the five-hundred-million-year-old Burgess Shale. But whether they have descendants or not, there is no doubt that extinction is the lot of most species on our planet. Roughly speaking, the lifetime of species seems to be a few million years, with a general rate of extinction in the fossil record being about 10 percent during what geologists call a stage—a period of time lasting between three and five million years.

To make this abstract discussion of extinctions a little more concrete, imagine a series of geological strata like that seen in the Grand Canyon. As you go from the lower to the higher strata, you move forward in time. Beginning at the bottom stratum, imagine climbing up until you find a stratum where the fossils of a particular organism are first seen. In general, that same fossil will be seen in the next few strata as you move up, but then it won't be seen anymore. The first stratum where a fossil is seen marks the appearance of that particular species on the world stage; the last stratum defines the point at which it becomes extinct.

It is probably worthwhile to talk a bit about the data that underlie the general statements I have made about extinctions. As we saw in chapter 2, fossils form when a plant or animal dies and is quickly removed from the biosphere. A long chemical process eventually replaces the calcium in the animal's hard parts with minerals, producing, in the end, an exact replica in stone of those hard parts. What we call the fossil record is simply the sum total of all fossils found and catalogued by scientists over the past few centuries. It constitutes the only hard knowledge we have of life and ecosystems in the past.

If you think for a while about the process of fossil formation, however, you will quickly see that when we look into the past, we are looking through a filter that distorts what was there. The fossil record is not a record of all past life, but of past life that survives in the form of fossils. There are several points that can be made about this fact.

First, most animals (and, for that matter, most species) are not represented in the fossil record. The way that different groups of organisms get represented in the fossil record varies, depending on how easily fossils form from the remains of members of the group. For marine invertebrates (like clams), as many as one out of a hundred species (not individuals) may be seen in the fossil record. For other groups, like land animals, that number may sink to one species in ten thousand, with the number going even lower for difficult groups like insects. Thus, statements about past life must necessarily involve an extrapolation from organisms that are in the record to organisms that aren't. This is not necessarily an easy extrapolation to make.

Second, the fossilization process tends to favor organisms that, in the normal course of affairs, will be removed quickly from the biosphere when they die. Thus, clams and mussels that live on (or even beneath) the ocean bottom are much more likely to be buried soon after death and make it into the fossil record. In point of fact, the great preponderance of entries in the fossil record are marine invertebrates—things like clams that lived in the shallow waters of the continental shelves.

Third, organisms that have prominent hard parts are much more likely to show up in the fossil record than those that do not. Fragile organisms (orchids, for example, or insects) are also much less likely to show up in the record, simply because they are less likely to remain intact after death.

The fourth aspect of the fossil record we need to consider is that it tends to favor marine organisms over those that live on land. The reason is simple—in the normal course of affairs, these organisms, whether clams, fish, or plankton, will sink to the bottom and be covered with sediment when they die. There is a much better chance, therefore, that their remains will be fossilized and found at a later date. The remains of land animals, on the other hand, are much more likely to remain at the surface, where they will decompose due to the action of scavengers, microbes, and oxygen.

It takes a singular event—a flood, for example—to remove their remains from the biosphere and begin the fossilization process.

Finally, the fossil record favors animals that are plentiful and have a large geographical distribution. There is nothing particularly mysterious about this bias in the record, either. If two animals have roughly equal chances of becoming fossilized when they die, then the species that has more animals—more entrants in the race—is more likely to succeed. Paleontologists in the future, in other words, are more likely to find twenty-first-century fossils of *Homo sapiens* than they are of some rare South American monkey, simply because there are more people around to serve as candidates for the fossilization process.

Taken together, this discussion of biases in the fossil record tells us that when we use it to look into the past we are likely to see a lot of marine invertebrates like clams and relatively few rare, fragile land-dwelling species. This fact has important implications for the discussion of extinctions, because it tells us that insects (the quintessential fragile land-dwelling animals) are likely to have a rather sparse fossil record. Since most extinctions taking place today are supposed to be among insects, this makes a direct comparison of past and present extinction rates difficult.

Although there seems to be a steady background rate of extinction in the fossil record, there are some events in which extinction rates become very high. When the dinosaurs disappeared 65 million years ago, for example, fully two-thirds of all other life-forms on the planet disappeared at the same time. This particular event, which scientists believe was triggered by the impact of a giant asteroid on the Earth, is an example of what paleontologists call a mass extinction. It was a period when great numbers of species suddenly disappeared, a truly catastrophic event.

The event that included the demise of the dinosaurs was not the biggest mass extinction we know about, nor was it the most recent. The biggest mass extinction occurred about 250 million years ago, at what geologists call the Permian-Triassic boundary. At this point in time, fully 80 percent of all species perished. The most recent mass extinction occurred about 13 million years ago. This one was relatively mild as mass extinctions go, with only about 30 percent of species disappearing. The cause of these mass extinctions is not really understood. Some scientists have suggested that they all correspond to meteorite impacts, while others argue that each is a distinct event, associated with its own special causes. However this debate turns out, the fact remains that the history of life on our planet is dotted with periods when great numbers of species became extinct. The question before us, then, can be put simply: are we now, because of human activity, entering another of these mass extinction events?

The answer to this question, of course, depends on comparing two numbers—the percent of species becoming extinct now and the number that vanished during some previous mass extinction. But from basic arithmetic, to get a single extinction rate, either current or past, you have to divide the number of species becoming extinct by the total number of species present. In the case of the fossil record, this number is obtained by dividing the number of species present in the fossil record whose last appearance occurs at a particular point in time by the total number of species present in the record at that time. To get the current rate, you have to divide the number of species becoming extinct today by the total number of species now present on the planet. Thus, to compare the current extinction rate to rates in the past, we actually need four numbers.

As we shall see, all four of these numbers are very poorly known. Thus, we have a situation where we understand how to answer the question in principle, but lack the data to make the comparison in fact.

This is not an unusual situation in the sciences. We often encounter situations where the true value of a particular number can't be determined exactly. In this case, we often find another number, a proxy, that we can calculate and that, we hope, gives us a reasonable estimate of what we really want. Provided we don't lose sight of what we're doing and provided we don't start assigning too much reality to the proxy, there is no particular problem with this approach. Difficulties can arise, however, both in public debates and policy discussions, when people ignore the caveats that, to scientists, are second nature.

A technical point: scientists generally look at a quantity like the number of species becoming extinct at a given time in a slightly different way from the way they look at a quantity like a rate of extinction. For example, the statement by an environmental advocate that "X species are becoming extinct every hour" and the statement by an ecologist that "In the next decade, Y percent of tropical songbirds will become extinct" are seen by scientists as being different in an important but subtle way. The point is that it often happens that a ratio like extinction rate (which is the number of species going extinct divided by the number present) can be determined more accurately than something like the number of species going extinct. You can see this by imagining a situation in which both the number of species going extinct and the total number of species are overestimated by a factor of two (i.e., both are thought to be twice as high as they really are). In this case you will get the correct extinction rate (because the errors in the two numbers cancel), but overestimate the number of extinctions.

Taking Accounts

With all of this in mind, let's look at how we go about estimating the total number of species that now share our planet with *Homo sapiens.* The obvious first step, of course, is to look at the scientific literature and add up all the species that have been discovered, described, and catalogued. When we do this, we find that there are roughly 1.8 million known species that share the Earth with us, the great majority of them insects.

But it is obvious that we haven't catalogued all the species in existence. The question comes down to how we are to estimate the number of species yet to be found—the species we know nothing about at this time. There are many ways to make such an estimate, with different techniques leading to widely divergent estimates of the number of species on Earth.

For example, we can look at the rate at which new species are being discovered. If we look at a group where we think most of the species are known (birds, for example), we see that early on the rate of discovery of new species is high. As more and more species become known, however, the rate of discovery drops, simply because there are fewer species left to discover. If we assume that the process of discovery is the same for insects as it is for birds, then we can estimate where we are in the discovery process for the former. When we do this, we come up with an estimate of between 5 and 7 million species on the planet.

Perhaps one of the biggest surprises in the study of insect species came in the late twentieth century, when scientists began investigating insect diversity in the tropics. One technique was to put a canopy over a single tree in the rain forest, then fumigate to kill all the insects present. A painstaking cataloguing process would then occur, while scientists studied each of the dead animals. The results were little short of astonishing. It was found that hundreds and even thousands of hitherto unknown species of insects lived in those trees, with every type of tree having an essentially unique insect community associated with it. I can remember one seminar, for example, in which the researcher reported that there were several hundred different species of beetles living in the fronds of a particular kind of palm tree. If you take these sorts of results for insect diversity in individual trees and multiply by the number of tree species in the rain forest, you come up with very high estimates of the number of species on the planet— as high as 30 million, in fact.

The point of this discussion is simple: no matter which estimate you take of the total number of species on the planet (and most scientists seem happy with an estimate between 8 and 10 million), the great majority of those species are unknown to us. This means that estimates of current extinction rates, like those of extinction rates in the past, are highly uncertain. This doesn't mean that we shouldn't make those estimates, but it does mean that we have to respect the uncertainties when we talk about the results.

Ecologists, unlike paleontologists, don't normally look at the fossil record, but approach the question by looking at the destruction of habitat. It is well known that the survival of any species depends on the survival of its habitat—the ecosystem in which it has evolved and on which it depends for its necessities. Cut down a forest to build a shopping mall, for example, and the plants and animals that lived in the forest must either move elsewhere or disappear. (Of course, other kinds of plants and animals will subsequently flourish in the shopping mall, but that is a different issue.) Cut down enough of the forest, and some of the species there will go extinct.

One of the tools that ecologists can use to estimate rates of extinction has to do with a well-known relationship between the number of species that a given ecosystem can support and the area of that system. The so-called species-area law says that the number of species, S, that can exist in an ecosystem of area A is given by the equation

$$S = \text{(some number) } A^n$$

where n is a number between 0.16 and 0.33.

A mathematical aside: an equation like the one above is called a power law. To see what it means, imagine that the number n is exactly equal to one. In this case, the equation would read

Number of species = (some number) times (the area of the ecosystem)

In this case, if we cut the size of the ecosystem in half (by cutting down half of the forest, for example), the number of species present would also be cut in half. The smaller forest, in other words, would have only half the number of species (mainly insects) that were present in the original forest. The fact that n is a number less than one means that if you cut down half the forest, you will have more than half of the original species left. If we take a typical value of n as 0.25, then this law tells us that if you eliminate 90 percent of a habitat, you lose half the species present. (This fact, incidentally, explains why conservationists fight so hard to maintain the integrity of large ecosystems, rather than let them be split up into smaller pieces. This is one case where the whole is actually greater than the sum of its parts.)

If you combine the high estimates of the number of insect species in tropical rain forests with the known rate of deforestation (which is a measure of the decrease in area of the tropical ecosystem), you can get a very high estimate for current extinction rates. If you've heard speakers talk about things like "a species disappearing every few minutes," they are basing the claim on this calculation.

It could be that this estimate is right, of course, but it is important to bear in mind that these are all *theoretical* extinctions. While the decrease in area of habitat is a reasonably firm number as these things go, the number of species is not. Thus, the species that are supposed to be disappearing have never been catalogued by scientists and, to put it bluntly, may never have existed in the first place.

In addition, there is some question about whether all of the species that would disappear according to the species-area law really go extinct or whether some of them just adapt to new habitats. According to Stanford University biologist Gretchen Daily, "More than half of Costa Rica's native bird species occur in largely deforested countryside habitats, together with similar fractions of mammals and butterflies." In this case, the "disappearance" of species from an ecosystem simply refers to the fact that they move or adapt to new surroundings, and not to any extinction event. (Nevertheless, the consensus among tropical ornithologists is that many of the original species will become extinct, even though hard evidence for this opinion is not yet available.)

Some scientists have been calling for a major effort to compile an accurate and exhaustive index of the Earth's species, and if this is ever done, we will have a better understanding of current extinction rates. Until then, however, honesty demands that we acknowledge the uncertainties in our estimates.

High, Low, or Uncertain?

So what are we to make of the claim that extinction rates now are higher than they have been since the disappearance of the dinosaurs—that human beings are causing another mass extinction event?

To my mind, this is probably the most uncertain of the claims made about extinction. It is indeed true that if you compare the estimates of current rates of extinction—even relatively conservative ones—with the rates of extinction seen in the fossil record, current rates are higher. However, you then have to look at the numbers that are being compared. The fossil record basically measures the rate of extinction of marine invertebrates in the past. The current extinction rate measures the theoretical rate of disappearance of insects right now.

> *If there was ever a case of comparing apples and oranges, this is it.*

So how do we proceed? It seems to me that there are two ways we could make meaningful comparisons of current and past extinction rates.

1) We could try to determine the rate at which habitat disappeared in the past and make the same kind of calculation ecologists make about current species, or
2) we could try to estimate extinction rates among current organisms that are likely to appear in future fossil records and compare that to what we see in the actual fossil record.

Both of these approaches have the enormous advantage of comparing apples with apples. Unfortunately, they are both difficult to do.

Let's start with the first approach. To define an ecosystem, we need to know something about plants—fragile land organisms unlikely to leave fossils behind. There is a fossil record of plants, of course, but it's pretty spotty. My paleontologist friends tell me that an event like the current deforestation of the Amazon Basin would probably not be seen in the fossil record. It would simply be invisible because of the filter through which we observe the past.

I talked to University of Chicago paleontologist Michael Foote about the second approach given above. With David Jablonski of the same institution, he did an interesting comparison—one that could only be done by the new wave of paleontologists who work as easily with computers as they do with the geologist's hammer. In fact, as Foote points out, "This work is a good example of how [our field] has changed."

The scientists began with two computerized databases that had been assembled for other purposes. One is the so-called Red List of endangered and vulnerable species. Looking only at mammals of North America, they were able to compile a list of mammals whose status was rated as "vulnerable" or worse. They then looked at another database called FAUNMAP, maintained by a consortium of paleontologists, which contains information on the fossil record of plants and animals in North America.

Foote and Jablonski then asked an interesting question. Suppose that all of the animals culled from the Red List became extinct tomorrow. What would some future paleontologist calculate for the extinction rate of North American mammals in the twenty-first century?

Look at it this way: if an animal has left a fossil record (i.e., if its presence is recorded in FAUNMAP), then future paleontologists would see that fossil record suddenly terminated—the classic signature of an extinction. If, however, that animal does not appear in the fossil record, then it will simply be invisible since, by hypothesis, it won't be around in the future to leave any fossils. The future paleontologist will therefore underestimate the number of extinctions at the start of the twenty-first century. In the case of North American mammals, Foote and Jablonski estimate that the actual extinction rate in their thought experiment will be double the one seen in the fossil record.

Today's scientists, looking backward, are in the same position as that hypothetical future paleontologist in the thought experiment. We see the extinction only of relatively widespread species that appear copiously in the record and, as a result, we think that extinction rates in the past were lower than they actually were. Add in the fact that what we really want is the extinction rate of insects and not that of mammals, and the underestimate becomes much worse.

So there you are. Like many issues regarding the current state of the planet, the question of human-caused extinctions gets messy when you get into the details. It's hard to get from this discussion to a bumper sticker, and I know that this will alarm some people. But the world isn't simple, and that's just something we have to live with.

For the record, I think it would be truly astonishing if something as far-reaching as the effect of human activity on the planet didn't drive some species to extinction. Whether the rate of extinction is truly unprecedented, however, is not so clear. I have to confess that I have this sneaking suspicion that animals like those four hundred species of beetles in the palm fronds have probably been becoming extinct at a high rate for hundreds of millions of years. After all, an animal so specialized that it can only survive on one part of one kind of tree is not a good bet to win the Darwinian sweepstakes. And, of course, since we have no idea how fast they became extinct in the past, we have no way of knowing whether their extinction rate is going up or down today.

In the end, it is clear that as human beings begin to manage the global ecosystem, we will have to make some hard decisions. Some of those decisions may well involve letting certain species become extinct. When those decisions are made, however, it should be with a clear-eyed knowledge of both the benefits and the costs of our actions, not on some vague (and basically unsubstantiated) rhetoric about human activities and mass extinction. In Mike Foote's words, "I'm all in favor of protecting biodiversity and habitat, but we don't have to throw away our credibility (as scientists) to do it."

Having made all of these caveats, I want to come back to the central question—since it is highly probable that human activities are causing some species on this planet to become extinct. What are we to do about this?

In a later chapter, I will lay out what I think is a sensible set of principles to guide us in the management of our planet. In the course of that discussion, I will look at various rationales that have been set forth to argue that it is in humanity's best interest to preserve as many species as possible. As a preliminary to that discussion, however, let me make a few points about what we have learned about extinction.

First, the great majority of the species that are supposed to be becoming extinct are highly specialized tropical insects—often insects so specialized that they can exist only in one part of one kind of tree. Faced with this fact, the reaction of many people is to ask, "So what?" For myself, I find it hard to believe that the disappearance of this sort of animal will have much effect on the future of the planet. To my mind, in fact, most of the practical arguments supporting species preservation are not very strong, a point to which I will return in chapter 14.

This leaves us with what is basically a philosophical or religious argument—the argument about stewardship. We should save species because it is the right thing to do (or, in a more stern version, because we have no right to destroy them). In my conversations with people on this issue, this is the argument that seems to carry the most weight. To many of my friends, preserving the environment has become almost a religious cause, and numbers like the ones dealt with in this chapter are simply irrelevant.

The problem with this point of view is that there is no way of deciding where to stop. It is manifestly impossible to save everything, and that means that hard choices still have to be made. It's difficult to make those choices if you have a belief that, in effect, assigns infinite value to every species on the planet. Like it or not, difficult decisions will have to be made, often, as in this case, in the face of insufficient scientific information.

9

The Greenhouse Effect and
Global Warming

As I pointed out in the first chapter, the issue of global warming is probably the most vexing and complex of the environmental issues facing us. At the most fundamental level, the science involved in projecting the amount of global warming is complicated, and understanding exactly what it is that climate scientists do is not easy. In addition, it is impossible to talk about global warming as a scientific issue alone. Any decisions made in this area will have profound effects on present and future generations, no matter what those decisions are. The problem is that the good or bad outcomes coming from those decisions will not be evident for a generation or more. This means that, to deal with the issue, political leaders have to inflict sacrifices on people today, but the benefits of those sacrifices will accrue to people who may not even be born for years to come. Finally, there is the simple fact that, although the outlines of a reliable theory of climate prediction may be emerging from today's science, our knowledge is not yet precise enough to allow us to say in any detail what the outcome of any particular policy decision is likely to be.

Put all of this together—the scientific uncertainty, the long timescale, the distant political payoff—and you have the makings of a truly intractable problem. In this chapter, I will explain why the scientific situation is as it is, speculate on when we might be able to expect higher levels of certainty in our predictions, and then talk about policy options.

A final word before we plunge into the details of climate models: in what follows, you will notice that I will be spending a lot of time talking about uncertainties. In part, this is because I feel that in the transition from scientific work to executive summary to press release, the uncertainties in our models of climate change have been largely ignored (although they have not been ignored by scientists). More importantly, though, I spent most of my research career building mathematical models (albeit models much simpler than those I'll be discussing below). That experience left me with an abiding respect for the pitfalls and uncertainties involved in trying to represent the world in this way. When I look at any model, the first thing I ask is "Where are the weak points; what could go wrong here?" You should be aware, then, that I'm not singling out climate models for special attention in this regard.

Weather changes from day to day—it may be sunny in Seattle today and raining in Phoenix, for example. Climate, on the other hand, involves weather patterns over long periods of time. Despite today's weather, we know that the general pattern calls for rainy weather in Seattle and warmer, drier weather in Phoenix. We say that Seattle has a moist, maritime climate, while Phoenix enjoys a dry, desert climate. When we talk about global climate, we are talking about the general weather patterns that prevail over the entire planet—patterns that can be expected to persist for long periods of time.

We can begin by making a couple of general statements about the Earth's energy balance. The driving factor as far as climate is concerned is the simple fact that more sunlight strikes the Earth in the tropics than at the poles. This is why the poles have ice caps and the tropics have palm trees. The Earth has to adjust itself to even out this imbalance. If the planet didn't rotate, the atmosphere would "boil"—warm air would rise at the tropics, move poleward as it cooled, and then sink at the poles, returning as cold air to the equator. In this case, winds would blow from the north in the Northern Hemisphere. But because the Earth rotates, this energy transfer becomes more complicated. The rotation causes the trade winds to blow from east to west in the tropics and prevailing westerlies to blow in the opposite direction farther north. The net effect of all this motion, however, is to transport heat poleward.

The great ocean currents move heat in the same way. Think of the Gulf Stream carrying warm water north along the coast of North America and returning to the tropics as cold water in the form of the Canaries current along the coast of southern Europe and Africa.

One of the key features of the Earth's climate is that it is always changing. Even in recent historical times, as we saw in chapter 5, we have gone through the Medieval Warm Period and the Little Ice Age, and on geological timescales, we have gone through even bigger swings associated with ice ages and Snowball Earth. So the news about global warming is not that the climate is changing—this statement could have been made at any time in the planet's history. The news is that, for the first time, this change may be associated primarily with human activity. The scientific issues involve such questions as how much of past and future changes, if any, can be ascribed to specific human activities, and how much will changes in those activities alter possible future changes in the climate. The policy issues revolve around the question of how to balance current costs of mitigation against uncertain future benefits.

The global warming issue starts with some very simple statements about the Earth's energy balance. Every second of every day, sunlight falls on the illuminated side of the Earth. This sunlight contains energy, as anyone who has laid out on a beach on a warm day can testify. (To be precise, an area of one square meter at the Earth's orbit receives 1,370 watts of energy from the sun.) Like all energy, the energy in sunlight can be neither created nor destroyed, but can only be changed from one form to another. This sunlight represents virtually all of the planet's energy budget.

What happens to that energy? Let's take a simple (but unrealistic) example to see. Suppose that the Earth had no clouds and that all the sunlight that came to its surface was absorbed. In that case, all the energy in the sunlight would be converted into heat and the planet would start to warm up.

But every object in the universe radiates heat—the warmer an object is, the more heat it radiates. Right now, for example, your body, which is at 98.6 degrees Fahrenheit, is radiating energy to its surroundings—one of several mechanisms the human body uses to get rid of excess heat. At the same time, your surroundings—a room at 70 degrees Fahrenheit, for example—are radiating heat energy toward you. Because your body is at a higher temperature than the walls, you radiate more energy than you absorb, and you feel comfortable. On a hot August afternoon in a city, your surroundings may well be radiating more energy than you are, and you may feel uncomfortably warm.

So, as the Earth absorbs energy from the sun, it heats up and radiates energy back into space. Eventually, the planet radiates as much energy as it absorbs and the whole system comes into equilibrium. You could say that the temperature stabilizes when the planet's energy books are balanced. The energy balance of the real Earth is a little more complicated than this, but the principle is the same. The planet will warm or cool until the amount of energy it radiates away is in balance with the energy coming in from the sun.

On the real Earth, some of the incoming sunlight—about 25 percent—is reflected back into space by clouds and never makes it to the surface, while another 5 percent is reflected by the surface itself. Another 25 percent is absorbed in the atmosphere, and the remaining 45 percent is absorbed by the surface. Both of the last two categories supply energy that serves to heat up the Earth, which then radiates energy back toward space.

The key part of this energy balance—the part that has relevance for the debate on global warming—has to do with what happens to the energy that is radiated up from the surface as it makes its way through the atmosphere. The energy coming in from the sun, whose surface temperature is about 5,000 degrees Celsius, is in the form of visible light. The Earth's atmosphere is largely transparent to visible light, so most of this energy comes right on through unless it is reflected. The surface of the Earth, however, is at a much lower temperature than the sun, so it emits its energy in the form of infrared radiation. (Infrared radiation is similar to visible light but has a longer wavelength. Your eye cannot detect it, but you can feel it as warmth on your skin.) Thus, the energy that comes into the atmosphere from the sun is in a slightly different form than the energy that leaves the surface and starts upward.

And therein lies the rub, because the atmosphere contains gases that are *not* transparent to infrared radiation. In this, the atmosphere acts a little like a glass greenhouse. The glass is transparent to visible light, so energy from the sun comes streaming in. It turns out, however, that glass absorbs a good deal of the outgoing infrared radiation, so the energy is trapped inside the greenhouse. The result: because of this and other effects, the greenhouse heats up until it is radiating enough energy to push through the absorption barrier and achieve balance. This is why you sometimes see global warming referred to as the greenhouse effect, although a real greenhouse achieves a good deal of its warming by keeping warm air from escaping—something the Earth doesn't do. You have had direct experience of the greenhouse effect if you've ever gotten into a car on a cold but sunny winter's day and found the inside to be pleasantly warm.

In the case of the Earth, the role of the absorbing glass is played by many different kinds of molecules in the atmosphere. Water vapor and methane (natural gas) are both "greenhouse gases," as are several others, but the greenhouse gas that is the center of debate is carbon dioxide. As the infrared radiation from the Earth's surface starts its journey back into space, some of it will be absorbed by carbon dioxide molecules in the atmosphere. Eventually, these molecules will reradiate the energy they have absorbed, but they are as likely to send it back toward the surface as outward into space. When this radiation reaches the surface, it is absorbed, and the surface heats up a little. The heating, of course, increases the amount of radiation being sent out toward space. As before, the surface heats up until enough energy is getting through the atmosphere to balance the planet's energy budget. You can think, then, of the greenhouse gases in the atmosphere as a kind of a blanket around the Earth—they keep the planet warm by slowing down the outward flow of energy.

Before we go on to a discussion of the global warming debate, I should make one point. There is absolutely nothing new about the greenhouse effect. The planet has always had one. We have already seen the importance of greenhouse gases in chapter 5, where we described how carbon dioxide, blown into the Earth's atmosphere by volcanoes, rescued the planet from its Snowball episodes. In fact, if it weren't for the greenhouse effect, the average temperature at the Earth's surface would be a frigid -20 degrees Celsius (about -68 degrees Fahrenheit). Without the greenhouse effect, in other words, the oceans would have frozen over long ago, and life would have been impossible.

The question on the world agenda today, then, is not whether human beings have created a greenhouse effect—that was here long before our ancestors came on the scene. The question is whether humans are, by their activities, altering the existing greenhouse effect enough to cause major changes in the Earth's climate.

There is absolutely no question that human activities are increasing the levels of carbon dioxide in the atmosphere. From a preindustrial concentration of about 280 parts per million, carbon dioxide levels in the atmosphere have risen steadily to a current level of about 360 parts per million. The cause of this increase is easy to see. Modern industrial economies run almost entirely on fossil fuels—coal, natural gas, and oil. The contributions from hydroelectric and nuclear power, while present, represent less than 15 percent of the energy budget in a country like the United States. All the rest of the energy to heat our homes, run our cars, and power our factories comes from burning fossil fuels.

Fossil fuels represent solar energy that has been stored in chemical bonds and buried underground by living things (plants in the case of coal, plankton in the case of petroleum). When we dig these substances up, we gain access to that energy. We obtain the energy by breaking the chemical bonds that formed millions of years ago, and the easiest way to do this is to allow the coal or oil to combine with oxygen—to burn it. This chemical reaction releases the energy we want, but it also creates carbon dioxide as a by-product. It is this carbon dioxide, released to the atmosphere, that accounts for the buildup we are observing today.

It is important to realize that carbon dioxide is not a "pollutant" in the normal sense of the word. That is, it is not something that is added to the atmosphere inadvertently while we're trying to get at something else, as happens when sulfur (a source of acid rain) is released when we burn coal to generate electricity. If you burn fossil fuels, you can only get the energy you want by letting the carbon atoms combine with oxygen—in other words, if you want the energy from fossil fuels, you get the carbon dioxide. Period.

Modeling the Climate

The next question, of course, concerns the effects of adding that carbon dioxide to the atmosphere. In general, you would expect the Earth to get warmer, just as you would expect your body to warm up if you added an extra blanket at night. But things aren't that simple. The planet is not just a blob of stuff that absorbs and radiates heat. The Earth responds to changes in temperature by altering its own structure. Let me give you a simple hypothetical example of something that was considered early in the greenhouse debate. If the temperature goes up, you would expect more water to evaporate from the oceans. With more water vapor in the air, you would expect more clouds to form, and some clouds reflect sunlight back into space. It is not impossible to suppose, therefore, that adding carbon dioxide could wind up cooling the planet. In point of fact, when scientists looked at this scenario in detail, they found that it didn't work—the formation of clouds depends on a lot more than just water vapor. Nevertheless, this kind of argument shows that a simple "increased carbon dioxide equals higher temperatures" line of reasoning won't work. If we want to understand the effects of the increased levels of carbon dioxide, we have to look at the Earth in all its complexity.

This is a difficult problem. The naturally occurring carbon dioxide in the atmosphere raises the Earth's temperature about 40 degrees Celsius (about 80 degrees Fahrenheit) above what it would be without any greenhouse trapping. On top of this very large normal effect, we are trying to make accurate predictions of warmings of a few degrees.

The tool of choice for scientists studying the climate are massive computer codes called Global Circulation Models (GCMs). These codes take into account literally thousands of different processes that happen on the real Earth. Is there more sea ice floating off of Antarctica? It will reflect more sunlight than the open water that it replaces would have. Is the boundary of the Sahara moving northward as a result of climate change? The open land will reflect more sunlight and absorb less water than land covered with vegetation. Are there more low clouds? They will reflect sunlight back into space. Are the clouds higher up? They will tend to trap heat.

All of these effects, and countless more, have to be accounted for in a GCM. These models are truly a monument to human ingenuity. Yet, given the complexity of the Earth, the relatively small effect we are trying to calculate, and the enormous political and economic repercussions that will follow from the results of the computer output, it is worth taking some time to think about exactly what the GCMs can do and, more importantly, what they can't.

To learn about this subject, I called on an old friend, Ray Pierrehumbert of the Department of Geophysical Sciences at the University of Chicago.

He is a prominent climate modeler—he was the first to include the effects of clouds in tracing the history of the atmosphere of Mars, for example. Pierrehumbert is a genial man, with a full beard and a permanent twinkle in his eye, but talking to him can be a little disconcerting. He is, not to put too fine a point on it, one of the smartest people I've ever encountered in a lifetime spent in academe. You get a sense that while he's listening to you, there is a dynamo spinning away inside his head, analyzing and dissecting everything that is being said.

What else can I tell you about Pierrehumbert? He plays a mean accordion—in fact, he has been known to bring his instrument in to serenade department parties.

But that day, over a lunch of cheese, sausage, and beer, we had another agenda on our minds. I had asked Pierrehumbert to bring me up to speed on the current state of the GCMs, a subject on which he knows a great deal. He started with the basics. "There are three segments to any GCM," he said, "the atmosphere, the oceans, and the coupling between them."

The roles of the atmosphere and ocean in the climate have already been discussed, but the coupling between them is important as well. Sometimes, as in the maintenance of the El Niño, the two interact directly, with prevailing winds pushing ocean currents and ocean currents maintaining the winds. But in terms of the greenhouse effect, there is a more important coupling. The ocean can both absorb carbon dioxide from the atmosphere and put it back in. In fact, most of the easily accessible carbon dioxide on our planet is sequestered in the oceans. You can think of the world's oceans as being a large carbonated fluid, something like your favorite soft drink. The delicate balance of carbon dioxide exchange between the oceans and the atmosphere depends on temperature (among other things) and is an important piece of any GCM. Much of the research effort in climate modeling today focuses on understanding this and other interactions between the oceans and the atmosphere.

"We have a strange vocabulary in climate modeling," Pierrehumbert continues. "We split the problem up into what we call dynamics and what we call physics." "Dynamics" describes the flow of air and water, while "physics" describes processes like the transfer of energy by radiation, the churning, heat-driven motion in clouds, and the like.

The basic process in any GCM is easy to describe, if difficult to carry out in practice. Scientists split the atmosphere up into a bunch of little boxes, and the equations that govern the flow of gases and energy are applied for each one. The computer then calculates how the gas in each box evolves for a fixed period of time, and at the end of that period, a grand accounting takes place. If the equations say that heat or radiation or material leaves a particular box, that energy or material is subtracted from its home location and added to the appropriate box elsewhere. Once this is done, the equations are applied to each box again, and the model steps forward one more interval into the future. In a typical GCM, this time interval is twenty minutes for calculations of the atmosphere and a few days for the ocean.

If this procedure were carried out with boxes the size of thimbles, everything would be fine. But our computers are limited, a fact that introduces one source of uncertainty in the results of GCM calculations. It turns out that if you make the boxes less than about 150 miles (200 km) on a side, with perhaps eleven boxes stacked vertically from the ground to the top of the atmosphere, calculations on future climates take so long that they become impractical.

As Pierrehumbert puts it, "These are the same computer programs that are used for long-term weather forecasting. To get a ten-day forecast, Europeans typically use a resolution of twenty kilometers, and the Japanese are making a serious effort to get down to one kilometer. If we used this kind of resolution in climate calculations, it would take a computer three months to go forward one year, and we have to predict the climate a century from now. We have to be able to run weather forecasts in a day, and climate predictions in a few months." As has been true since computers were first developed in the mid-twentieth century, one of the things that limits our ability to predict both weather and climate is the speed and capacity of our best computers.

There is an interesting issue involving the use of computers in the climate debate. Pierrehumbert points out that we are still programming in FORTRAN, a computer language that was developed in the 1960s. In this, the climate modelers are in the same position as air traffic controllers in the United States were not too long ago. They were using computers and computer codes that were decades out of date. It was not at all unusual for secretaries at airports to have computers that were vastly superior to the ones being used by the men and women directing the flow of airplanes. This is only to be expected in an endeavor where complexities are piled onto complexities, and where going back to square one can be enormously difficult and expensive. Nevertheless, Pierrehumbert and his colleagues are initiating a transition to modern computing methods and software, which should result in a rapid improvement of our ability to predict future climates and the impact of human activities on them.

But there are other, more physical problems as well, each of which introduces its own kind of uncertainty into climate calculations. It is to these that we now turn.

Sources of Uncertainty

Any mathematical model of any natural system, whether it runs for months on a massive computer or is scribbled on the back of an envelope, is an attempt to capture, in a precise logical structure, the essential characteristics and behavior of some feature of nature. Here's a simple example to show what I mean: suppose you throw a baseball up into the air and want to describe its ascent and fall. A real baseball is a complicated thing—it has a rough leather covering, stitches in a pattern on its surface, irregularities in its internal structure, and so on. It will be moving through air that is also inhomogeneous as it climbs up above a lumpy and irregular Earth. If we tried to take all of these details into full account, we could probably generate a computer program that rivaled GCMs in complexity.

But we don't have to do this. Instead, we build a mathematical model of the baseball—in effect, we create in our minds an alternative universe where things are simpler. In that alternative universe, the baseball is a smooth, homogenous sphere, and the Earth is the same. We get rid of the air, so the ball moves through a vacuum. With all of the details stripped away, we have a simple model of the moving baseball. In exchange for throwing away the details of the baseball, we get a problem that can be solved by any student in a freshman physics class.

When a mathematical model is created, then, there are two questions that can be asked: (1) how important were the details you threw away to get to your model, and (2) does the imaginary world you created in your mind really correspond to the world we live in? The second question is called the problem of validation, and we'll return to it in a moment.

The question of the importance of details is where the modeler's art becomes important, because what is "detail" and what is "essential" depends on what you are trying to do. In the example of the baseball, treating the baseball as a perfect sphere is fine if you want to know the time it will be in the air, but not good enough if you want to calculate the resistance offered to the ball by the air.

And this is where it makes a difference that the calculated greenhouse effect is small compared to the natural one. Because we are calculating a small effect, we have to "keep the stitches on the baseball" in GCMs in order to get accurate results. That is what makes the models so complicated.

But there has been enormous progress in these models over the last decade. Because I got involved in writing articles on this subject for magazines like *Smithsonian* and *Time,* I have had a box seat as this work has gone on. Ten years ago, for example, the models had great difficulty handling two of the major determinants of climate: ocean currents and clouds. The standard model of the ocean portrayed it as a kind of swamp—water a few meters deep, capable of storing heat but little else. Today, most of the details of the ocean and its coupling to the atmosphere can be included in calculations. This is an enormous step forward, but different models, differing from each other in technical details, still produce widely different predictions of warming in the next century.

In fact, the Intergovernmental Panel on Climate Change (IPCC), an international body in which thousands of climate modelers around the world participate, routinely reports its results as a spread of possible temperatures. Their standard result, which is widely quoted, is that temperatures by 2100 will be between 1.4 and 5.8 degrees Celsius (2 and 11 degrees Fahrenheit) warmer than they are now, with a best guess around 2.5 degrees Celsius (5 degrees Fahrenheit). This spread has nothing to do with statistics, but is simply an indication of how much the models differ in their predictions.

The difference between 1 and 6 degrees Celsius is a big one. A 1-degree Celsius warming would be about the same as what has occurred since 1850 and could be handled with a few more air conditioners. A 6-degree Celsius warming would be something like what happened after the last Ice Age and could well require (among other things) building dikes to protect coastal cities from rising sea levels. (I will return to the subject of unlikely events with drastic consequences later in this chapter.)

When I asked Ray Pierrehumbert what causes the greatest uncertainty in climate-warming predictions, he answered in a single word: "Clouds."

Clouds are pretty to look at, but they are enormously complex physical systems. Clouds are made up of water droplets (or, sometimes, ice crystals). Their life is a constant battle between evaporation and condensation in those droplets. If evaporation wins, the cloud vanishes. If condensation wins, it grows. The details of how clouds wax and wane are extremely complex and still something of a puzzle to modern science. Yet clouds play a major role in the development of climate, since they both reflect sunlight and trap heat near the Earth. More importantly, they tend to be a few miles or a few tens of miles across and therefore are very difficult to treat in a computer program that has boxes 150 miles on a side.

"In terms of the kinds of things that are important in making political decisions," says Pierrehumbert, "there will not be much of an improvement in the uncertainty in GCMs over the next decade. The physics is just too complicated."

This is an important statement, because it is precisely during the next decade that we will have to make important decisions about the global-warming issue. It means that we will have to make those decisions in the face of clear scientific uncertainty, about both the scope of the warming and the consequences of our activities.

The Real and the Modeled World

Before going on to consider the policy implications of this situation, however, I would like to take a little time to talk about the problem of validation of the GCMs, because it illustrates something that I consider to be a serious difficulty in the current debate. There is no question that in the model worlds inside the GCM computers, there is a clear greenhouse effect and global warming. The question that has to be answered, however, is whether the world that exists inside the computer is the same as the world we live in.

There are many ways you can go about answering this sort of question. You can look at various pieces of the GCM and test them against nature.

You can, for example, check your calculations of cloud formation or ground cover against field studies. You can ask whether the GCM predicts the right outgoing radiation by comparing its results against satellite measurements. You can ask if the model gets some of the gross features of the Earth climate right—does it, for example, get summer and winter temperature difference right; does it get the right relation between temperatures at the poles and the tropics? In these sorts of studies, the models seem to do reasonably well. In addition, when the models are used to explain the climate during past ice ages, when carbon dioxide levels were very different from what they are now, they get a lot of the main features of those climates right. All these results give us some confidence in the models.

But when I think about validation, I always ask what I consider to be the gold standard question: "If I gave you the climate in 1900, could your GCM predict the twentieth century?"

Ten years ago, the answer to this question was an unambiguous no. The answer is still no, but the situation is a bit more ambiguous. The twentieth century was quite complicated as far as climate goes. In the beginning of the century, there wasn't much of an increase in carbon dioxide. Then the warming that might have been expected as carbon dioxide levels increased from the 1950s to the 1970s was largely canceled out by the increase in aerosols in the atmosphere (think smog and haze), which reflected more sunlight back into space. These aerosols are largely a result of the same human-engineered processes that produce the carbon dioxide. (To be honest, I have to say that this realization is largely a matter of hindsight—only a few people were thinking about aerosol cooling in the 1980s.) In any case, the end result is that it is only in the last decade or so that scientists have been able to see clear evidence for the predicted warming.

When I asked Pierrehumbert my gold standard question, he thought for a long time before answering. "It is possible to adjust the parameters in some of the new models to fit the twentieth century." We cannot yet, in other words, reproduce the climate of the past century from first principles, but we can make the most advanced versions of the GCM consistent with the twentieth-century climate record. While the current situation represents a significant advance over what came before, it still falls short of giving us the kind of confidence in the models that would be needed to convince skeptics that it is time to impose costly changes on our economy.

This is one reason why the debate over global warming has become so polarized.

You might wonder what it means for a scientist to "adjust" a model—the term sounds as if it involves fudging the results, but it is actually a well-accepted part of the modeler's world. There is, in any mathematical model, a certain amount of play, of wiggle room in the parameters. In other words, it is always possible to change the parameters of the model a little bit without violating any known fact about the system, and this means that the outcomes—the predictions—of the model can be adjusted a bit as well. This, of course, adds one more level of uncertainty to those predictions. To take a hypothetical example, suppose you wanted to include the effects of high clouds, which are mainly made from ice crystals, in your calculations. The "right" way to proceed would be to take the known properties of water and ice, fold in the atmospheric conditions, and come up with a prediction that might say something like "At this point in time, this percent of the sky in this region will be covered with these sorts of clouds." The model would then take this result and use it to calculate things like levels of reflected sunlight and the trapping of infrared radiation—essential ingredients in the final predictions of temperature.

But suppose you don't know enough about cloud formation to carry out this process? Another way to proceed would be to adjust the percent of cloud cover (in this hypothetical example) to match past climate data. Suppose, for example, that the amount of cloud cover in a given situation could vary from 20 to 40 percent and still be consistent with known data about cloud behavior, and if you could show a cloud cover of 32 percent, it would bring your predictions into line with past climate data. In the language of modelers, you would then "adjust" the parameter representing cloud cover to 32 percent and use that number in making predictions of future climate. In a much more sophisticated and multifaceted way, advanced GCMs can be adjusted to "predict" the twentieth century.

As I said, this is a huge step forward for the modelers, and to my mind it represents a reasonable way to deal with uncertainty. It suffers, however, from one important deficiency. The problem is that when you adjust parameters to get a desired outcome, you have, in essence, given up any attempt to link that outcome to basic physical processes. If you then take your adjusted model and try to predict the future, you are flying blind. You are assuming that whatever it was that made your parameters work for the twentieth century will also make them work for the twenty-first. This may be a good assumption, but then again, it may not be. You just can't know.

Having said all this, however, I have to stress one important fact. Although there may be some wiggle room left in the GCM, the amount is shrinking all the time. It now looks pretty certain that if we adjust the parameters in a GCM to match the twentieth century, that model will predict some level of warming for the twenty-first. In other words, the amount of uncertainty in the model predictions no longer allows for a world unchanged by the continued use of fossil fuels.

The Case of the Missing Carbon

Just to give you a taste of how complicated the debate on global warming can be, let me tell you about one of the nagging problems that has bedeviled the field in the past: the problem of the missing carbon.

We know roughly how much coal, oil, and natural gas is burned in the world every year, so we can get a pretty good estimate of how much carbon humans are putting into the atmosphere. The last decade for which we have detailed data on the fate of this carbon is the 1980s, when human activity put about 5.4 billion metric tons of carbon into the world's atmosphere. For reference, a billion metric tons of carbon in the familiar form of graphite (pencil lead) would cover an area the size of an average city to a depth of a few feet. A good deal of this carbon, as pointed out above, will go into the ocean, and some of what is left will be incorporated into the world's plants, particularly in forests. Carbon taken into the molecules of substances like wood or incorporated into minerals is said to be sequestered. You would expect, then, when you add up all the carbon taken into the ocean and into living tissue and soil and subtract it from the total being put into the air, you should have the net amount of carbon accumulating in the atmosphere in the form of carbon dioxide.

The problem is that when you carry out this exercise, the books don't balance. About half of the carbon we put into the air every year seems to vanish without a trace. This is obviously a source of concern, because in formulating policy options, understanding processes that can remove carbon from the atmosphere—they're called carbon sinks—is obviously important. Global warming can only be caused by carbon dioxide in the atmosphere, so what matters is the amount of carbon put into the air minus the amount sequestered. More than one international negotiation, for example, has foundered on the question of how to assign credit to countries for sinks such as forests.

The search for missing carbon has a long history. There was a time when scientists thought that it was being taken up by living things in the ocean, with some deposited on the ocean floor as organic detritus. My daughters, preteens at the time, delighted in calling this the fish poop theory. Today, our models suggest that the carbon being pulled out of the air is roughly divided between carbon going into the ocean and carbon going into land sources.

Most scientists are confident that we understand how carbon is taken into the ocean, but the corresponding process on land remains a problem. Factors such as deforestation, changing land-use patterns, forest growth, and so on complicate our models. The models seem to be telling us that the missing carbon is being taken up by the Northern Hemisphere continents— that it is presumably being sequestered by forests. This makes sense when we think about the United States, where land that was farmed a century ago in New England has reverted to forest today. When we look at forest inventories worldwide, however, we see no evidence for the carbon that is supposed to be there. In the words of Steven Wofsy of Harvard University, "If our forests are sequestering billions of tons of carbon annually, why can't we find it?"

Today, research on this question focuses on those parts of forests that are not normally included in forest inventories, such as deadwood, woody debris on the forest floor, and scrub timber encroaching on open fields. Since forest inventories are maintained primarily to support logging operations, these potential carbon sinks are not counted. Scientists estimate, in fact, that in the United States fully 75 percent of the carbon in forests is not counted in the traditional inventories.

It may be that we are finally approaching a reliable way of calculating the effect of carbon sinks on the global carbon budget. If this is so, then we can look forward to a time when it is possible to make reliable estimates of the net carbon being added to the atmosphere by each nation, a necessary first step toward international controls and regulation.

The Question of Risk

One of the most difficult aspects of dealing with the subject of global warming is the wide range of outcomes consistent with our current models. As pointed out earlier, these range from relatively low cost to extremely expensive adjustments, depending on where in the predicted spread of one to six degrees of warming you think the planet will wind up. This means that we have to deal with the concept of risk, always a difficult subject. Worst-case scenarios, however unlikely, make great copy for newspapers and TV commentators, and hence tend to dominate public discussion. In addition, research on human risk perception shows that human beings tend to overestimate the probability of catastrophic events—to think, in other words, that these events are much more likely to occur than they actually are. All of these issues enter the picture when we talk about global warming.

The uncertainties in the science of the GCMs (mainly those associated with clouds) could end up driving the predicted heating either up or down—we won't know until we solve the problem. There are, however, other, more speculative (and probably more unlikely) processes that could have even larger effects. For example, some scientists have argued that there is evidence in climate records contained in ice cores that when warming reaches a certain level, there are other, as yet poorly understood, processes that will push the planet into a new climate regime, perhaps by altering the Great Conveyor Belt discussed in chapter 5. If this is true, the warming could end up being worse than the prediction of a 6-degree Celsius warming that is now the IPCC upper limit.

On the other hand, as we shall see in subsequent chapters, people are talking about developing technologies that would allow us to pull carbon dioxide out of the atmosphere or at least prevent it from getting there after burning fossil fuels. We might, for example, remove carbon dioxide from smokestack gases at power plants, more or less in the way we now remove pollutants. This carbon would then be sequestered in a sludge. Alternatively, there are a number of schemes that involve finding ways to make organisms like ocean plankton or bacteria remove the carbon dioxide from the air. If any of these technologies are developed, you can imagine a future in which fossil fuels continue to be burned, but in which there is no attendant warming.

For the record, I think that pure "End of the World" and "Technology Will Fix It" scenarios are equally unlikely and that the future will involve some mix of warming and technological mitigation.

Although the uncertainties in the climate future may seem daunting, we deal with these same sorts of extremes all the time—for example, when we get behind the wheels of our cars. It is easy to imagine gruesome worst-case scenarios resulting from a car trip. We normally don't dwell on them, however, because we know they are unlikely. On the other hand, good drivers pay attention and exercise caution because they know that they can't assume everything will be fine regardless of their own behavior. In just the same way, we have to find a middle road in thinking about the problem of global warming. It is to that issue that we now turn.

Given all of the complexity surrounding climate change, how can we evaluate the situation? Every scientist will have a different answer to this question, probably depending more on questions of innate levels of optimism or pessimism than anything else. For the record, I'll tell you how I come down on some central questions, since these answers have an obvious impact on my suggestions as to how we should manage the Earth as a whole.

First, based on the best available data, there is little doubt that since 1850 or so, the planet has warmed up and that this warming has accelerated in the past few decades. There are a handful of outstanding issues here, involving some discrepancies between satellite and surface temperature measurements, but they seem to be getting resolved in favor of warming. I have to remind you, however, that the Earth has warmed and cooled continuously in the past, so the simple fact of warming is not a particular reason for concern. The real question is whether human activities are responsible for the warming.

This is a much more complex issue. For one thing, we know that we have been coming out of the Little Ice Age since the middle of the nineteenth century—long before any human activity could have had much of an effect. For another, there are always scientists around who play the role of devil's advocate and delight in pointing out other processes that could lead to warming, but that aren't included in the GCMs. One of these, the possible slight warming of the sun, remains a subject of debate.

On the other hand, as climate models get better, it becomes possible to study what are called signatures of human-driven warming—general characteristics of climate trends that seem to be associated with increased levels of greenhouse gases. For example, regional warming (and cooling) patterns seem to be in line with the predictions of the GCMs. While the evidence remains messy, the understanding of these signatures seems to be getting better.

The bottom line, then, is that while evidence for human agency in global warming is getting better, we don't yet have an open-and-shut case. If the warming is due to global trends beyond our control, then all we can do is think about adapting to higher temperatures. If, on the other hand, the warming is due to human activity, then we can exercise some control over the climate's future.

My own guess (and I have to stress that this is a guess) is that we will find that at least some, and perhaps even most, of the observed warming can be attributed to the use of fossil fuels by human beings. In a sense, this is good news, because it would give us some control over our future.

When I come to the bottom line in this sort of argument, however, I realize that there are really two interrelated questions involved: (1) how much faith are we to have in the GCMs, and (2) how much are we willing to sacrifice today to avert the global warming predicted by the models? For myself, I don't yet have enough faith in the GCMs to give up my full-size car and house in the suburbs to move to a cramped apartment near a bus line. And frankly, I don't think many people who profess a lot more confidence in the global-warming predictions than I do are ready to make a comparable sacrifice either.

The No Regrets Option

So now you can see the problem of global warming in all its gory details. We have enormously complex GCMs that are improving by the year, along with many difficult problems still to be solved before we can make accurate predictions about the outcome of any policy option we might choose. Furthermore, it appears unlikely that science will improve enough in the next decade, the time during which action must be taken, to give policy makers a magic bullet that will solve our problems. What is to be done?

You can think of the possible responses to global warming as falling on a spectrum between two extremes—"let her rip" and "shut everything down." The first of these is often called the business-as-usual approach, and it proceeds on the assumption that we will be able to get through somehow, regardless of what happens to the climate. The second sees the threat to the planet as so great that activities that add carbon to the atmosphere should be stopped immediately, no matter what the cost. (To be fair, I have never met anyone who takes either of these positions, although I have met a few who come pretty close to each extreme.)

The "right" way to settle such an issue would involve some sort of cost-benefit analysis. You should be able to say something like "If we burn this much fossil fuel, then the resulting changes in climate will cost X dollars, while the cost of not burning the fuel is Y dollars." Factor in the things on which it is hard to put monetary value (the well-being of a national park, for example, or the hardship on the working poor of shutting down transportation systems), and you have a basis for making rational decisions. Unfortunately, as we have seen, the GCMs aren't yet good enough to do this nor are they likely to give us the information we need in the near future. We simply don't have the data we need to make choices, even if we could agree on criteria by which those choices should be made.

This situation is actually not unfamiliar. We have an uncertain but identified risk (global warming) with a large cost attached to it. As I pointed out above, you face the same sort of situation every time you drive your car. There is a risk that you will be involved in a serious accident, and that risk has a high cost. The way you handle the risk attendant on driving is to buy insurance, and I would suggest that the process of paying insurance premiums is a good model for policy decisions about global warming. No one spends all of his or her income on insurance premiums, but only an idiot buys no insurance at all. Looked at this way, the global warming debate comes down to a discussion of how much insurance we want to buy—how much of our wealth we want to spend on premiums. This approach has the advantage of turning a rancorous debate, in which one side accuses the other of either destroying the planet or ruining the economy, into a more normal political discussion.

In this spirit, let me tell you about a policy option that makes sense to me. It's called the "no regrets option," and I have heard it discussed at scientific meetings for a long time. The argument goes like this: suppose, just for the sake of argument, that the whole global warming thing turns out to be a fiction, a nonproblem like Y2K (remember that one?). Are there things we could do now to mitigate warming that we would not regret having done, even if the warming never materialized? If so, then these are the policies we should be pursuing.

When you look at the industrialized economies in this way, it's not hard to find candidates for the no regrets option. Some of these are small things—would anybody be upset if we developed more efficient home appliances, which did the same job as the current ones with lower energy use? Would we lose anything if we built homes out of two-by-six lumber instead of two-by-fours, so that we could insulate them better? Or if we developed better and more efficient ways to recycle materials? Thirty years from now, I think people would be glad to have these technologies in place, regardless of what the GCMs of the day are saying.

There are, of course, larger projects that I think would fit into the option as well. Right now all of the economies in the industrialized world are run almost entirely on fossil fuels—coal, oil, and natural gas. There is no particular reason why this situation should continue into the future. Alternative technologies are already available—fuel cells (either hydrogen or methanol), solar panels, and wind energy are all either close to being or already economically competitive with fossil fuels. The United States actually has a huge wind resource, centered in the High Plains states from North Dakota to Colorado, and farmers and ranchers are discovering that they can turn the wind that blows over their land into a cash crop by leasing windmill sites. Thus, the development of wind energy might even have, as a by-product, the preservation of the family farm—a social bonus.

The United States has actually gone through two great changes of its main energy source in the last couple of centuries—from wood to coal and then from coal to oil. Such changes take about thirty years to complete, as old equipment wears out and is replaced. I suspect that thirty years from now people will be very happy to be driving electric or fuel-cell cars, using energy provided by solar panels in Arizona and windmills in North Dakota. Big energy companies and automobile manufacturers are already getting involved in these sorts of projects. Some judicious support of research efforts, along with tax incentives and some regulatory changes, are probably all it would take right now to get us to where we want to go, and this seems to me to be a small cost regardless of where the global-warming debate takes us.

Let me put this another way: how much would you be willing to pay to stand on the stern of the last oil tanker to leave the Middle East, waving good-bye as you go? I, for one, would pay a lot, which means that developing renewable energy sources is an excellent candidate for a no regrets policy.

Having said all of this, I have to add that many of my colleagues (including, I should point out, Ray Pierrehumbert) think that I should be a lot more worried than I am. Maybe so, but right now the uncertainties in the predictions are too big, and the level of political invective in what should be a purely scientific debate too high, for me to come to any other conclusion.

In the end, then, each of us has to look at this enormously complex issue, factor in his or her assessment of and comfort with risk, and decide what steps need to be taken. Each of us needs to respect the fact that others may disagree with us and that disagreement is not a sign of moral failure. Then, together, we may be able to deal with the problem in spite of the uncertainties.

III

The New Environmental
Toolbox

10

Genomics

What is life?

One of the greatest achievements of nineteenth- and twentieth-century science was finding the answer to this question. And make no mistake about it, the question is difficult. Aristotle, for example, speculated that magnets might, in some sense, be alive because they moved. For a long time, people believed that there was some kind of "life force"—élan vital—that made living things different from the nonliving. Had this turned out to be the case, then the study of living systems would constitute a completely separate branch of science, with no connection to mundane chemistry and physics. Over time, however, we have come to realize that there is nothing singular about the mechanisms of living things. The hydrogen atoms in your body, for example, are the same hydrogen atoms that exist in the ocean, in the sun, and, for that matter, in the most distant galaxy. Living things just aren't all that different from everything else. In fact, the great truth that scientists have learned is that

Life is based on chemistry.

No matter how complex or how simple, whether we're talking about a single-celled bacterium moving toward food or a trillion-celled human being unraveling a problem in general relativity, at bottom life is nothing but molecules coming together and breaking up. It's only chemistry—there's nothing all that unusual about it. The chemistry of life may be very complicated, and we are certainly a long way from understanding it all. Nevertheless, we do know enough to be certain that, while it may be complex, life is not fundamentally mysterious.

Actually, this understanding of life has been around for at least fifty years. What's new is that today we not only know that life is based on chemistry, but we are starting to figure out how that chemistry works. We are, in effect, "getting under the hood" of living things, figuring out what makes them tick, beginning to tinker with the nuts and bolts of things. This is one of the great advances in science that will, in time, form the basis for the new relation between humans and nature. In this chapter, I will outline the basic understandings we have gained, and in the next, I will discuss some of the social and political problems that grow from this new knowledge, particularly in the area of genetic modification.

I was a graduate student at Stanford in the late 1960s, back when silicon valley was in the process of becoming Silicon Valley. I remember the start-up companies housed, quite literally, in garages, the sense of infinite possibility that seemed to permeate the air. Today, of course, we recognize that what was being born on the San Francisco peninsula back then was nothing less than the information revolution, which has already changed the world almost beyond recognition. Because of this experience, I know what it's like to be around a revolution when it's just getting started.

Whenever I go to the high-tech corridor that sprawls along Interstate 270 north of Washington, D.C., one of the centers of modern biotechnology, I get that same old feeling. Things are happening there—new ways of understanding the world, new ways of manipulating life, new technologies. This is a revolution that's just starting, a revolution that could leave what happened in Silicon Valley all those years ago in the dust. In fact, you can already see it happening in today's headlines—think of genetically modified plants and animals, clones, stem cells, and the rest. One of our great problems in the coming decades will be to bring our social, religious, and political systems up to speed in this new technological renaissance.

The Mechanisms of Life

If life is based on chemistry, then that chemistry is based on a molecule called DNA. The best way to picture the DNA molecule is to imagine a ladder. The sides of the ladder are made from two alternating molecules—one is a sugar called deoxyribose and the other molecule is made up of one phosphorus and four oxygen atoms. The deoxyribose is what the D in DNA stands for (the N and A stand for nucleic acid, which says that the DNA molecule is found in the nucleus of the cell).

Note that the term "molecule" refers to any collection of two or more atoms bound together, so there is no problem in having one molecule (like DNA) made from other molecules.

It's actually the rungs of the ladder that are interesting. They are made up of two molecules called bases, one sticking out from each side of the ladder and linking together to form the rung. The four bases, called adenosine, cytosine, guanine, and thymine, are usually represented by the first letters in their names, A, C, G, and T. The shape of the molecules is such that A and T can fit together, as can C and G. A, however, cannot fit to C, nor can T to G. It is only if the molecules fit that a rung can be constructed on the DNA ladder. Thus, there are only four possible rungs:

AT
TA
CG
GC

Run up and down one side of the ladder and you will find a sequence of these molecules—ATCCGATTTA, and so on, each linked to its partner to form a rung.

I should point out that once the DNA ladder is constructed like this, with the rungs made from base pairs, you can give it a twist in your mind to convert it into the familiar double helix shape.

It is the sequence of bases along the DNA molecule that contains the information needed to run the chemistry of a cell. In order to understand how this is done, you have to know one more thing about the kinds of molecules that occur in living things. It may be true that they are made from the same kinds of atoms as ordinary stuff and that they obey the same laws of physics and chemistry. Nevertheless, they tend to be large, complex things, made up of hundreds and even thousands of atoms.

The key point is that molecules don't form bonds with each other—only atoms can do that. In each large, convoluted molecule, then, there are places—scientists call them active sites—where the atoms can form bonds with active sites in other molecules. Think of these sites as little Velcro patches on the molecules. If you bring the two molecules together so that their Velcro patches line up, they will stick together. If the orientation is wrong, they won't bond. As with everything else in the molecular world, what matters is shape and orientation.

The chances of two complex molecules coming together in just the right orientation to align their Velcro patches is pretty small, so nature has provided another way to run chemical reactions in cells. Here's an analogy that may help you understand how this system works. Imagine that you have two long ropes, each one with a Velcro patch. If you just threw the two ropes on top of each other, the chances of those Velcro patches lining up is essentially zero. Suppose, however, that instead of throwing the two ropes into a random heap, you picked them up, aligned the Velcro patches, and stuck them together. In this case, the ropes would bond together every time, because you had oriented the ropes properly.

There is a suite of molecules that do for other molecules in your cells what you did for the ropes in this example—they bring molecules together so that they can interact. The molecules that perform this function are called enzymes. An enzyme is a molecule that facilitates an interaction between two other molecules, but does not itself take part in the reaction. In the above example, you acted as an enzyme—you made sure the ropes stuck together, but did not become stuck to a rope yourself. An enzyme is like a real estate agent—he or she brings the buyer and seller together and makes the transfer of a house possible, but does not actually buy a house.

The molecules that act as enzymes in living cells are called proteins. Each different protein molecule has a shape that allows it to facilitate one and only one chemical reaction. In effect, the protein grabs each of the two interacting molecules and makes sure they line up properly, allowing them to do whatever it is they do. Thus, if we know what proteins are present in a cell, we can tell what that cell's chemistry will be. It is, of course, this chemistry that gives each organism its distinctive characteristics. It is also, ultimately, the ability to understand and control proteins that will give human beings one of the tools they need to become the managers of the planet.

The best way to think about proteins, which are the workhorses of living systems, is to imagine that they are constructed like a bead necklace. You make the necklace by choosing one bead at a time and sliding it onto the string. If you have a large assortment of colors from which to choose each bead, there are a virtually unlimited number of different necklaces you can put together. The "beads" from which proteins are made are smaller molecules called amino acids, and there are twenty different "colors" of amino acids found in living things. Once the amino acids are strung together to make a particular protein molecule, the electrical forces between the amino acids cause the molecule to fold up into a unique shape. The interacting molecules then fit into the hills and valleys on the protein's surface, and this is how the orientation function is performed. The shape of the protein determines its function as an enzyme, and this shape is determined by the sequence of its amino acids. Thus, to run a specific chemical reaction in a cell, we need a protein with a specific sequence of amino acids—a specific sequence of "beads" on its "string."

And this is where DNA comes in. The message contained in some segments of the DNA molecules gives the instructions for the sequence of amino acids—the sequence of beads on the string—for making a specific protein. We call the part of the DNA molecule that carries these instructions a gene. The gene is the basic unit of inheritance, the structure that allows the characteristics of the parent to be passed on to the child.

Before I start into an explanation about how a cell turns DNA code into proteins, I should warn you that the process will seem complicated. In fact, I often tell my students that learning about the details of molecular processes in living systems is a little like reading a Russian novel. On each page, there is a perfectly understandable interaction between the characters, but by the time you get to page 1,437, you can't remember if Pyotr Pyotrovich is the cousin or nephew of Alexei Alexeivich. In the same way, individual reactions in cells are simple, but by the time you have followed through to the end of a process, there are so many molecules involved that you have trouble keeping track of them all.

With this in mind, let's talk about how a gene works: each group of three rungs on the DNA ladder determines, through the action of a suite of intermediate molecules called RNA (ribonucleic acid), what the next amino acid will be in a protein under construction. Tell me what those three rungs on the DNA ladder are—ATG, for example—and I'll tell you which amino acid will be the next bead on the string—the amino acid methionine, in this case.

RNA is just like DNA except that (1) it has a slightly different sugar on the side of its ladder, (2) it's only half of the ladder—the bases are not attached to other bases, so there are no rungs, just bases sticking out looking for something to attach to—and (3) a base called uracil (U) is substituted for T, but it has the same shape and does the same things.

The first step in the process of making a protein involves "unzipping" the DNA ladder (think of sawing through the rungs) and copying the information in the sequence of bases in the gene onto a short RNA molecule. The copying process basically involves having enzymes assemble the RNA by hooking up one base at a time, with the base on the new RNA molecule being the one that fits onto the exposed base on the unzipped DNA. In the end, the sequence of bases on the half ladder bears the same relation to the sequence of bases on the DNA, as a photograph does to a negative. Thus, if there is a C on the DNA there will be a G on the RNA being assembled, if there is a T on the DNA there will be an A on the RNA, and so on.

The point is that the RNA molecule is only as long as the gene and so is much shorter than a DNA molecule. It can get out of the nucleus of the cell, which the DNA cannot do, and carry that information out into the body of the cell, where the work of assembling the protein can take place. For this reason, this version of the molecule is called messenger RNA, or mRNA for short.

It may help you to think of the cell as a chemical factory. The DNA is in the central office and contains the instructions needed to assemble whatever final product the factory needs to make. In this analogy, the mRNA carries that information from the central office out onto the factory floor.

Once there, another kind of RNA takes over. Called transfer RNA, or tRNA, it has three unattached bases on one end and a place at the other end that fits one and only one amino acid. In effect, tRNA is the bridge molecule that takes us from the DNA code to the amino acid in a protein. The bases on the tRNA attach to three unattached bases in the mRNA—again, it's just a matter of which molecules fit together. If the three bases in the mRNA are CCG, for example, then the tRNA molecule that has GGC on its top will be the one that hooks on. And since each type of tRNA molecule attaches to one and only one amino acid, the next amino acid in the chain is determined. In this example, the tRNA attaches to the amino acid called glycine, so that will be the next bead on the protein string. In the end, this process adds one amino acid at a time to the growing chain until the protein is finished.

The bottom line of all this molecular machinery: the sequences of bases on the DNA ladder determines the sequence of amino acids in a protein and therefore the shape of that protein. This, in turn, determines what chemical reaction that protein will run.

All living things on Earth use this same basic machinery to turn the information on a gene into a protein that runs one chemical reaction in a cell. Everything in this process depends on nothing but molecules fitting together—it's all a giant jigsaw puzzle. No molecule in this sequence knows or cares where another molecule came from. The only question is whether they fit together or not.

Perhaps the most astonishing discovery we have made in this molecular age is this:

All living things on Earth share the same genetic code and use the same mechanism to assemble proteins from information in DNA.

It's hard to exaggerate the importance of this finding. At one level, given what we know about evolution, it is not too surprising. Since we're all descended from that first cell, it makes sense that we would all have the same basic kind of chemical machinery inside of us. Because of this fact, it is possible to splice DNA from one organism into the DNA of another and have the resulting molecule function as a single unit. This is what makes genetic engineering possible and provides human beings with an important new tool in dealing with nature.

The most important consequence of the fact that all living things share the same genetic code is that the machinery we've just described is completely insensitive as to whether the DNA it is copying and turning into proteins was originally a part of the cell or not. When you get the flu, for example, a virus co-opts your cell's machinery by introducing its own RNA into the system. The rest of the machinery doesn't care whether that RNA came from the cell's own DNA or not—it just churns away, producing copies of the original virus instead of the proteins your cell needs. The HIV virus, which causes AIDS, actually injects extra DNA into the DNA in certain cells in the immune system, and the machinery goes ahead and produces more viruses from that blueprint.

Obviously, from the point of view of basic cellular operation, the important part of the DNA molecule is the part that codes for a specific protein—the part we call a gene. It is something of a surprise to learn, then, that in human DNA only about 5 percent of the DNA is made up of genes. The remainder used to be called junk DNA, but that is surely a misnomer. Some of this DNA must be involved in turning genes on and off, and some may provide protection against viruses, but the complete understanding of the other 95 percent remains to be discovered.

There is one more detail about genes in multicelled organisms like humans that I should mention. It turns out that the base-pair sequence in our genes is interrupted by stretches of nonsense DNA called introns. It would be as if I started to say "w-da-babbada-babbada-babbada-ord," and it came out like that. There are pieces of the cell's machinery that know to cut out the nonsense syllable and produce a final product that says "word." This same machinery can also produce several different proteins from each gene, simply by leaving out certain pieces of the code. On average, each of our approximately thirty thousand genes codes for about three different proteins, so our cells need about one hundred thousand different kinds of enzymes to run their reactions.

The Human Genome Project

There are some events in history that stand out as landmarks, even though they are part of a continuous process of discovery. Usually they acquire a kind of symbolic importance—think of the first lunar landing, for example, or the invention of the electric light. In 2000, another such landmark event occurred with the announcement of the first assembly of the entire human genome.

A word of explanation: human DNA consists of about three billion base pairs, or rungs on the DNA ladder. A complete reading out of the bases from the first to the last would be called a complete sequence of human DNA. This sequence, which would contain about as much information as three sets (not volumes!) of *Encyclopedia Britannica,* would contain everything there is to know about the chemical working of the human cell.

In fact, having all of that information wrapped up in a single superlong molecule would be very unwieldy. Instead, the DNA is broken up into shorter segments (think of using scissors to cut a long string into shorter pieces), which are wrapped around a core made of proteins. These shorter segments of DNA are called chromosomes, because in the nineteenth century, scientists looking at cells under microscopes discovered that they would take up a red dye and appear to be colored. In humans, there are twenty-three pairs of chromosomes in all cells except the egg and sperm, which have half that number. There is no particular connection between the number of chromosomes an organism has and its complexity. Mosquitoes have only six chromosomes, for example, while dogs have seventy-eight. It's really just a question of how the DNA is packaged.

A complete reading of the human genome, then, would begin with the first base pair on chromosome one and finish with the last base pair on chromosome twenty-three. In fact, this is the way that scientists first thought of the task—later called the Human Genome Project (HGP)—when it was proposed in the 1980s. As originally conceived, the HGP would have been an enormously tedious (and very expensive) undertaking. It would, quite literally, have taken armies of biologists working for decades to finish the job. Because of this, there was a fair amount of resistance to the notion of undertaking the HGP in the beginning. I can remember one graduate student telling me over a late-night beer, "I don't want my life's work to be sequencing from base pair one hundred thousand to base pair two hundred thousand on chromosome fourteen."

But, as always happens, human ingenuity proved to be more than a match for the problem. Automated sequencing machines—essentially black boxes that take in DNA at one end and print out a sequence at the other—were quickly developed. In addition, a totally new technique was invented for the project, a technique that harnessed the power of the computer. It was the brainchild of J. Craig Venter, former president of Celera Genomics, a private company founded for the express purpose of sequencing all of human DNA.

Here's how it works: a stretch of DNA is cut into relatively small lengths—perhaps a few thousand base pairs. Each small segment is fed into an automated sequencing machine, and these machines, working in parallel, read the two ends of their short segments quickly. At this point, extremely sophisticated computer programs take the output of the sequencing machines and reassemble the information, producing a sequence of the entire long stretch of DNA. Think of this process as being analogous to reading a book by tearing the pages of many copies into little bits, having the opening and closing sentence of each bit read by a different person, and then having a computer reassemble the multiple readings into a copy of the original book.

Incidentally, the central role of computers in this process explains why many scientists prefer to talk about the bio*informatic* revolution, rather than about biotechnology. Without the computers and the sophisticated programming developed for them, that graduate student I talked to would probably still be slogging his way through chromosome fourteen.

What was celebrated in the summer of 2000 was called the first assembly of the human genome. The shredding and reading and piecing back together had all been done, and, for the first time, human beings were in a position to obey the Socratic dictum "Know thyself," at least at the molecular level.

There were, as was only to be expected, some surprises. The number of genes on the human genome turned out to be smaller than had been predicted—about thirty thousand instead of the predicted hundred thousand. From what I hear, this caused some consternation among lawyers for various biotech firms, who were worried that contracts calling for the delivery of one hundred thousand genes would be challenged in court. There was also a great to-do in the science press about how this somehow diminished human beings in the animal world—after all, we had only thirty thousand genes.

This part was rather silly. As we shall see, our new understanding of complex systems teaches us that you can get very complex behavior out of systems that have only a few relatively simple parts, so there is no particular correlation between the number of genes and the complexity of an organism.

The way that a cell runs is much more complicated than the quick overview given above, of course, but for the first time, human beings are starting to understand its basic mechanisms. It's hard to overestimate the impact that this will have. The first time we achieved this sort of understanding of the physical world was in the time of Isaac Newton, and it led, eventually, to the industrial and information revolutions. One way to make this point is to note that whenever a new understanding of nature is achieved, sooner or later someone will come along and find a way to make money out of it, no matter how abstruse and useless the knowledge may seem at first. For example, modern information technology, from the computer to the cell phone, comes from research into the structure of matter and the development of quantum mechanics that was carried out in the early twentieth century. And just as no one in 1930 could have predicted that his or her work would eventually lead to portable CD players and the Global Positioning Satellite (GPS) system, no one today can really see all the things that are going to come from the ongoing revolution in biotechnology.

Nevertheless, we can see some broad outlines of where things are going in the near future. Right now, many scientists are spending their time trying to mine the information contained in the human genome. Since the whole point of cellular machinery is to produce proteins that act as enzymes, an entirely new field of science called proteomics is springing up. The basic thrust of this field is to concentrate on what the proteins do in the cell, rather than how they are coded in the DNA. This is a complicated undertaking, because although the basic idea of how an enzyme works is simple, there are often dozens of enzymes that interact with each other to produce a specific effect.

One researcher likened the problem of using the results of the HGP to reading the *Encyclopedia Britannica*. It's one thing to have the text, quite another to understand what it means. Since 2000, we have had the basic text of human chemistry in front of us. Our job now is to figure out what it all means. In the case of human DNA, researchers are concentrating on genes that have a known connection to human disease, since that is where the most immediate payoff will be.

The first results are likely to be improved tests for specific diseases and conditions. The basic idea is that if we know what proteins are supposed to be present, we can devise specific tests to determine their presence or absence. We can expect to see new kinds of medications based on this new knowledge. These improvements won't lead to anything particularly dramatic in your experience of medicine, except that the tests will be more accurate and the drugs will work better.

The real changes will start when we get to what William Haseltine, a leading researcher and the CEO of Human Genome Sciences, calls regenerative medicine. Based on our new knowledge of the basic mechanisms of living systems, Haseltine sees progress coming in three distinct stages. First, in the near future, we will begin to use our knowledge of human genes and the proteins they produce to stimulate the body to make repairs—to heal skin lesions, for example, or stimulate the growth of blood vessels in the heart. He expects to see these sorts of changes on a timescale as short as five years.

This will be followed, in his view, by advances that will come from understanding how cells use molecules to signal to each other—how they decide, for example, that they will be muscle or skin or liver cells. As we begin to understand how this process works, we will be able to grow new organs for transplant—organs that will not be rejected by the body. Hasletine thinks that this may start to happen on a timescale as short as ten years.

Finally, in the very long term, we will learn how to reset the genetic clocks in our cells. This will allow us to reseed the body with new organs as old ones wear out. When this happens, an individual will no longer have a single age, but may have many ages, corresponding to a series of renewals.

Given the promise for improved human health, it's pretty clear that our new knowledge of the genome will be exploited rapidly, even pushed into controversial areas like stem cell research and therapeutic cloning. From the point of view of this book, however, this work won't have much impact on the relationship between humans and nature. Although we will live longer, healthier lives—a good thing—it does not necessarily follow that we will change our method of dealing with our planet. The real impact of genomic knowledge will come not from our ability to improve medicine, but from our ability to modify the DNA of all living things—the process that we now call genetic engineering or genetic modification. Whether this process will (or should) ever be applied to human beings is a complex question beyond the scope of this book. But it can be—indeed, already is being—applied to other organisms, such as food crops. It is this ability to change the very nature of life itself that will play a key role in our future, and to which we now turn.

11

Genetic Modification

Ten thousand years ago our ancestors made a fateful, though unconscious, decision. They decided that they would no longer be content to subsist on the food that nature, left to itself, provided for them. Instead, as we saw in chapter 3, they began raising crops and domesticating animals, thereby providing more food than was obtainable by simple hunting and gathering. This is an example of the human ability to overcome limitations imposed by nature. It is a profoundly human trait—some might say *the* profoundly human trait—and agriculture is just one example of it.

Today, thanks to the new understanding of genomics outlined in the previous chapter, we are on the brink of making a move that will be potentially more far-reaching than even the development of agriculture. We are rapidly approaching a point where we will no longer have to be content to work with the DNA that nature has provided for living things on this planet. Instead, just as our ancestors did thousands of years ago, we will be able to change our surroundings to suit our needs, except that this time we will be intervening directly into the genetic makeup of other living things. What I will try to show in this chapter is that we are already entering a phase of history in which we will modify the genetic structure of our planet, just as our ancestors modified the agricultural structure.

Let me start our foray into this new world by looking at a very simple example, the modern computer-assisted design of pharmaceuticals. As we saw in the previous chapter, all living things—including human beings—are basically complex chemical systems, with the business of life being carried out by billions of molecules coming together and breaking apart in a finely orchestrated dance. When you take a medicine, even something simple like aspirin, you are introducing a new molecule into this mix. This new molecule is supposed to either do something your own molecules aren't doing or stop your molecules (or those of an invading organism) from doing something they shouldn't be doing. They accomplish this task by attaching themselves to molecules in your system, a task that requires that they be the right shape to fit onto those molecules.

For example, when you have an infection, your doctor prescribes an antibiotic. The infection is caused by bacteria, which are multiplying in your body and producing toxins that make you sick. Part of the growth of the bacteria involves expanding their cell walls, a process you can picture as something like building a wall from Lego blocks. With Lego blocks, you fit one block onto another until you have a structure. In the same way, a bacterium fits one molecule onto another to expand its cell wall. The antibiotic fits onto these molecules, acting roughly like a wad of chewing gum stuck to a piece of Lego. The bacterium, unable to expand its cell wall as it grows, literally explodes. With the bacteria gone, you start to feel better. (And, since human cells do not have walls, the antibiotic has no effect on you.)

Like everything in the molecular world, the antibiotic works because it has the right shape—it's a simple lock-and-key situation. In fact, the same is true of every drug or medicine you might take—somewhere in your cells, the molecule in your medicine fits onto a molecule in your body. In the molecular world, shape is everything.

Historically, we have developed new drugs by looking at molecules that have already been produced by the process of natural selection. Folk medicines use these molecules in the form in which they appear in nature; modern pharmaceuticals use them as a base from which to build improved molecules—molecules that can pass more easily through the walls of the intestine, for example, or that have a shape that allows them to perform their job better. But the goal is always the same—find a molecule that is the right shape to do the job you want done. For a pharmaceutical industry based on this model (as it was throughout the twentieth century), nature is a vast storehouse of molecules, and by rummaging through that storehouse, we can usually find something that has the shape we want.

Over the past couple of decades, however, an alternative approach to drug development has started to come to the fore. Instead of looking only at available molecules, scientists start by looking at the job to be done and asking what shape molecule is needed to perform a certain task. A chemist then builds a molecule specifically to perform that task. (In principle, the building could be done from scratch, but it is usually more convenient to start with a molecule already available in the industrial database.)

For example, in one step in the life cycle of the HIV virus, some long protein molecules produced by the viral DNA have to be trimmed and snipped off. The molecule that does this job looks like a doughnut—the protein fits into the "hole," and then the molecule contracts and performs the cutting task. Scientists looking at this process realized that one way to block the virus was to find another molecule that would fit into the "hole" and keep the protein out. It was like fitting a cork into a bottle.

It happened that I was visiting Upjohn Pharmaceutical Company in Kalamazoo, Michigan, while a drug like this was being developed. I was invited into a strange room where I put on special glasses that allowed me to see in three dimensions, and then *I walked into the molecule!* A computer was projecting light into the room, and you felt you were literally standing inside the molecule, see where the atoms were, and, most importantly, see their shapes. You could see, for example, that moving the sulfur atom 30 degrees to the left would allow the fit to be better, that removing carbon would make things even better, and so on. In fact, it was this kind of computer-assisted design that produced most of the suite of drugs that allows AIDS victims to manage their disease these days. Conscious design of the drug molecules represents a radical new departure in medicine.

Here's an analogy that may help you visualize these two different ways of producing pharmaceuticals: suppose you have a bottle, and you want to keep some liquid inside it. This is the "task." One way to do the job would be to find a large pile of corks and rummage through it until you found one that fit the bottle. This represents the traditional approach of the pharmaceutical companies. Alternatively, you could measure the opening in the bottle and design a cork to fit it. This is what computer-assisted design is all about. As our knowledge of the "bottles" and "corks" increases, you can expect to see this approach come to dominate the industry.

Incidentally, as I shall point out later, this new approach makes me very skeptical when I hear environmentalists argue that we should save the rain forest (or some other ecosystem) because it might contain a cure for cancer. From the point of view of drug development, the rain forest is basically a huge pile of corks through which we can rummage. But if we start designing our own corks, we won't need the rain forest anymore. It is best, I think, to avoid an argument that could so easily be eclipsed by technological advances.

Folk medicine, in which molecules from nature are used in essentially unaltered form, can be thought of as a kind of hunter-gatherer strategy for drugs. Twentieth-century pharmaceutical production, in which molecules supplied by nature are altered to perform their function more efficiently, can be thought of as analogous to agriculture. But computer-assisted design represents a new departure, perhaps the first example in medicine of what I have called the second step. It involves nothing less than replacing what nature offers by something designed specifically for our needs.

But as revolutionary as this process is, it is really little more than a baby step on the road to the future. The real revolution will come not from modifying or designing molecules to be used as drugs, but by modifying the molecule that carries the code for the working molecules in the cells of other living things.

Genetic Modification

The DNA molecule lies at the heart of every living thing, carrying the codes that govern the construction of the enzymes that run the basic chemistry of life. The complex machinery of the cell turns this code into real molecules. And, as we saw in chapter 10, all living things use the same code and the same machinery. It is this unity of life at the molecular level that makes genetic modification possible.

Start with a stretch of DNA, with the base pairs making the rungs of the ladder. There are a set of molecules, called restriction enzymes, that will cut the DNA where a particular set of bases—ATT, for example—occurs.

Think of these enzymes as being something like a pair of scissors, reaching in and snipping three of the rungs on the DNA ladder. The net effect is that the DNA at the point of the cut has three unattached bases—ATT in our example. Think of these bases as forming a kind of Velcro strip at the end of the severed molecule. If another stretch of DNA happens to be around, and if the new stretch has a matching set of bases—TAA, in this case—then the two Velcro strips can make contact and attach. In effect, what you have done is to take two strands of DNA from different sources and tie them together to make a new DNA molecule.

If you can perform this sort of splicing once, you can do it as many times as you like. This means that you can cut out a stretch of DNA from one source and insert it into another DNA molecule, an operation that requires two splicings (one at each end of the DNA being spliced). And because all living things share the same genetic code, *it doesn't matter whether the DNA being spliced comes from the same organism or not*. The machinery in the cell will treat the new code just as it treats the original code—it will, in other words, proceed to turn that code into proteins and chemical reactions in the cell.

And this, in turn, means that we can take a gene from one organism, graft it onto the DNA of another, and expect it to work as if it had been in its new molecular home all the time. As far as the host organism is concerned, once the gene is there it is just like any other gene—the machinery goes to work and produces the protein it codes for. It is this fact that allows a virus to co-opt the cell's machinery to make new versions of itself. But what viruses can do, humans can do as well and, as a matter of fact, have been doing for decades. When we do it, the process is called genetic engineering or genetic modification.

For example, people who suffer from diabetes need a steady supply of insulin to manage their disease. It used to be that this insulin was obtained from the pancreatic glands of slaughtered pigs. This led, occasionally, to allergic reactions in some patients. Starting in the 1980s, however, the gene for human insulin was engineered into a common bacterium, *E. coli*. (Technically, the gene was not inserted directly into the bacterial DNA, but was in the form of a small, self-contained loop known as a plasmid.) The bacteria were then kept warm, well fed, and happy, and as they grew and multiplied, they produced human insulin as their cellular machinery dealt with their new gene. Virtually all insulin used in the treatment of diabetes is now produced in this way. This marks a major (though little-known) triumph for genetic engineering.

The biggest use of genetic engineering in the United States today, however, is not to be found in medicine, but in agriculture. Since herbicide-resistant soybeans were first introduced in 1996, use of genetically modified plants has skyrocketed. Between 2001 and 2002, for example, the amount of genetically modified agricultural planting increased by 13 percent. Today, almost three-quarters of the soybeans, a third of the corn, and almost a quarter of the cotton grown in this country has been genetically modified.

Why is this being done? There are several reasons. One use of genetic engineering in agriculture involves the attempt to control insect pests. Since the dawn of agriculture, predation by insects has been a problem. It remains so today, as any home gardener can tell you. Insects like the European corn borer and the boll weevil can cause significant damage to crops (and drive up prices for consumers). In modern agriculture, pests are controlled by successive spraying of insecticides.

But the process of natural selection has endowed most plants with their own internally generated pesticides. In particular, an obscure bacterium known as *Bacillus thuingiensis* has a gene that produces a very effective one. The gene for this pesticide, called Bt, was first transferred into corn, and then into many other commercial crops. If this gene is present in the plant, an attacking corn borer will be killed as soon as it starts to feed, before it can do much damage to the plant. This, in essence, is how a genetically modified plant operates. A useful gene from one species is inserted into the DNA of another, and the host acquires the advantages that natural selection has conferred on the donor.

Despite the furor that surrounds genetically modified plants, there is a major environmental benefit associated with Bt. Because the pesticide is in the plant itself, a good deal less pesticide has to be sprayed on fields to control insects. For example, in the Midwest, a field of sweet corn might have to be sprayed twelve to twenty-five times in a growing season to control the corn borer. Runoff from this spraying enters the biosphere and can cause harm to animals such as birds and fish. Scientists estimate that the use of Bt corn and cotton cuts the exposure of animals to sprayed pesticides by two-thirds and the exposure of humans to residues of sprayed pesticides by one-third.

Another emerging use of genetic engineering may actually play an even more important role in environmental policy. Like insects, weeds are another enemy of agricultural plants. When you are raising a crop for food, you don't want precious nutrients and water taken up by plants that are useless to you. Traditional weed control involves cultivation—driving a tractor through the field and turning over the soil in the rows between crop plants—and perhaps spraying specialized herbicides. A farmer in the American Midwest may have to cultivate his fields many times to control invading weeds. In the process, he disturbs the topsoil and makes it easier for that soil to be eroded by wind and rain.

But suppose you could genetically engineer your crops so that they were not affected by a particular herbicide. In that case, you could spray your fields with that herbicide, secure in the knowledge that it would kill all the weeds and leave your crop plants alone. This is exactly what Monsanto has done with a line of crops that are called Roundup Ready. Roundup is the name of a widely used agricultural herbicide, and Roundup Ready plants have been engineered to be resistant to it. Using these seeds, a farmer can engage in what is known as no-till agriculture. A field is sprayed with Roundup before planting to get rid of existing plants, then sprayed as needed to control weeds while the crop is growing. The result: the soil is never plowed up and erosion is reduced to a minimum.

These innovations represent a first wave of genetic modification, and I'll turn to the question of what the near and far futures might be later in this chapter. First, however, we need to talk about the controversies that surround the use of genetically modified organisms, or GMOs, in the industrialized world, particularly in Europe.

At one level, it can be argued that almost everything human beings have eaten for the last ten thousand years has been genetically modified. When early farmers chose certain types of wheat to plant, they were, over time, changing the gene pool of their plants by the process of artificial selection. When breeders crossed animal or plant strains to produce hybrids, they were mixing DNA from two different sources in the hope that the process would produce a more useful offspring. In fact, every domesticated animal and plant used by humans is the result of this kind of DNA mixing and has been for millennia.

But traditional breeding techniques are clumsy, a kind of blunderbuss approach to getting what you want. When you cross two strains of corn, for example, you have no idea what the DNA of the offspring is or what kinds of combinations you are producing. Genetic engineering is a more targeted approach—you identify the gene that has a useful property and transfer only that to the host.

Looked at this way, you might wonder why there should be any controversy about genetic engineering. Given the nature of the molecular mechanisms of cells, why should it matter where a particular gene comes from? If a gene from a flounder can make a plant more frost resistant (as it can), or if a gene from a bacterium can help a plant deal with insect pests, what difference does it make to the consumer?

Actually, there is a serious debate about the wisdom of widespread genetic engineering in agriculture. Some of this is not scientific in nature. First, however, I would like to address the scientific aspect of the debate, which centers around several issues: the possible spread of introduced genes from domesticated to wild plants, the possible toxic effects of wind-blown pollen on other insects, and, probably most important, the safety of the food supply for humans.

In traditional molecular biology, genes can only move vertically through populations. That is, they can be passed from parent to offspring, but not from one species to another. Over the past decade, however, scientists have started to realize that this rule may not always be honored. In fact, there is now good evidence that what scientists call "gene flow" occurs. In general, the flow is between cultivated plants and their wild relatives (between sunflowers grown in a farm field and nearby wild sunflowers, for example). As of this writing, there is no firmly documented case of a gene from a genetically modified plant being transferred into the wild, but it's hard to imagine that it won't happen sooner or later. In this respect, the use of GM plants with herbicide resistance may turn out to be analogous to the development of antibiotic resistance in bacteria. If enough weeds acquire Roundup resistance, for example, we may have to move to a new herbicide, just as pharmaceutical companies are developing new antibiotics in response to developing resistance to the ones already in use.

Oddly enough, the question of gene diffusion may lead to the reintroduction of a type of genetically engineered plant containing what are called terminator genes. From the point of view of agricultural companies, one of the problems with genetically engineered crops is that once they are developed, there is nothing to keep farmers from producing their own seeds (rather than buying them from the developer). In the late 1990s, Monsanto began developing improved crops that contained genes that would prevent the plant from propagating once the first growth season was over—these contained the so-called terminator genes. Some critics perceived this as an attempt to exploit poor Third World farmers, who would be forced to buy seed for each planting, so the program was dropped. Because of the concern over gene diffusion, however, this particular technology may be making a comeback, since the terminator gene would limit the ability of other engineered genes to spread. If it turns out that the diffusion of genes to wild plants constitutes a problem, this is one possible technology that could be used to manage it. It would have little effect in countries like the United States, where farmers customarily buy new seed every season anyway.

The concern over possible harmful effects of genetically engineered crops on insects blew up in 1999, when laboratory experiments showed that monarch butterfly caterpillars would die if they ate milkweed leaves that had been coated with Bt corn pollen. The monarch butterfly quickly became the poster child of the anti-GMO movement and was present at countless protests. By 2001, however, no fewer than six field studies had shown that pollen from actual crops used in the United States did not contain enough Bt to harm the caterpillars. In effect, not enough pollen containing Bt got onto plants normally eaten by the caterpillars to harm them. In the words of entomologist Mark Sears of the University of Guelph in Ontario, "The impact of this technology on monarchs is negligible." Again, as of this writing, there are no examples of collateral damage to other species from genetically engineered crops.

Having said this, I should add that these studies do not close this issue—they merely show that in this case the spread of pesticides by wind-blown pollen is not a problem. As we begin to use genetic modification on a larger scale, I expect that we will acquire a base of experience and knowledge that will let us know when to be careful and when we can proceed without risk in this area.

The main controversy over genetically modified plants, however, does not concern the environment, but the safety of the food supply. There is a concern (understandable, I think) that this new and untested kind of food may hold as yet unanticipated dangers for an unsuspecting public.

In the United States, there are three agencies that deal with the introduction of new kinds of crops, including genetically modified organisms (GMOs.) The Department of Agriculture (USDA) is concerned with whether the new crop is safe to grow, the Food and Drug Administration (FDA) with whether the crop is safe to eat, and the Environmental Protection Agency (EPA) with whether it is safe for the environment. Genetic engineering started in the 1980s, with approval of genetically modified foods by the FDA, and large-scale planting of GMOs started in the mid-1990s. Consequently, the United States has over a decade of experience in regulating these new crops. This, I believe, is one important difference between the United States and Europe, which only recently came to the business of GMO regulation.

The basic standard used by the FDA is one of "reasonable expectation of no harm," with approval depending on a determination of whether or not a new crop is "substantially equivalent" to foods already on the market. This makes sense, because, in the words of biologist Susan Harlander of BIOrational Consultants, "Food is not safe." Every food contains molecules that could potentially cause harm, and the reasonable way to judge new foods, whether genetically modified or not, is to ask whether they are as safe or safer than what is already in the food supply. As often happens in this sort of situation, it is impossible for any food to be absolutely safe; foods, like life, carry risk, even foods that humans have consumed for thousands of years.

The aspect of safety issues connected with GMOs that has received the most scientific attention involves allergies. At the molecular level, an allergy represents the response of a person's immune system to the introduction of a specific protein molecule. The result can be anything from an annoyance, as when proteins in airborne pollen grains trigger a mild case of hay fever, to life threatening, as when people allergic to proteins in bee venom die when their throat tissues swell up and suffocate them after they have been stung. A certain segment of the population will be allergic to certain kinds of food—shellfish, peanuts, monosodium glutamate (the so-called Chinese restaurant syndrome), and so on. People who have these allergies quickly learn to avoid the foods that bring them on.

But what if the gene that produces an allergy-inducing protein is transferred to another crop? What if, in other words, someone who knows not to eat shellfish unwittingly encounters the same protein in corn or potatoes? You can see that there is a serious potential for harm here.

One line of approach, of course, is simply to make sure that no known allergens are present in a new food, or, if they are present, to make sure that the product is labeled. Since there is an extensive list of these molecules, this is probably the most important (and most obvious) step to be taken. It already constitutes an important part of the American response to GMOs.

After this, scientists can look at the proteins that actually exist in the genetically engineered plant and try to determine if they could be harmful. Unfortunately, there is no general test for proteins that might trigger allergic reactions and no animal model that can help. Therefore, organizations like the FDA use some generalized tests to screen foods. For example, they can see if a particular protein is digested in humans (most allergens aren't). They can also see if the proteins are stable against heating (most allergens are). Eliminating proteins that fail to pass these (and similar) tests, then, becomes another line of defense against unknown allergens. (I have to point out that many common foods—the peanut is a good example—couldn't pass the screening of GMOs in the United States.)

In the end, the question of safety comes down to whether or not some as yet unknown allergen will be introduced into the food supply by genetic modification. At the molecular level, the question is whether by introducing a new set of proteins into the average diet we are taking a chance of causing harm. In my view, this is a pretty small risk, although a real one. It is, however, a risk that we as a society seem to have been willing to take in the past.

For example, when tomatoes were first introduced into Europe, they were widely regarded as poisonous. I have heard of an annual ceremony somewhere in New Jersey that re-creates an eighteenth-century event in which a man publicly ate a tomato to disprove the myth. (I understand that the point of the ceremony is for people in the crowd to yell, "Don't do it!")

In our own time, many new food products have been introduced in the United States, each representing an influx of unfamiliar proteins. Some of my readers are probably old enough to remember when kiwi fruit was unknown in the United States. Importing kiwi fruit introduced a whole slew of new genes and proteins into the food supply. For most human beings in the Northern Hemisphere, the molecules in kiwi fruit represented a new set of inputs into the body's chemical system. As far as I can tell, however, there wasn't much in the way of adverse pubic health effects following the introduction. Since there were a lot more unfamiliar genes and proteins brought in by both tomatoes and kiwi fruit than there would be in any GMO, we can take some reassurance from this experience.

Perhaps the most visible debate over GMOs is taking place in Europe, where strict limits are being put on imported agricultural supplies. I have to admit to some bemusement at watching my European colleagues get terribly upset over the risks of genetically modified soybeans while at the same time tolerating smoking—a much greater health hazard by any measure—at levels way beyond what is tolerated in America. Some of my more cynical friends have suggested that the whole thing is just a way of protecting inefficient French farmers. I will leave this question to others more knowledgeable in these areas.

In any case, the strategy that seems to be developing in the European Union is that imports of foodstuffs have to be certified to contain less than 1 percent GMOs. I don't know how to say this delicately, so I'll be blunt. This seems to be just about the dumbest, most bureaucratic way of approaching the problem I can imagine. Let me explain why.

Since any harm associated with GMOs will come from the molecules contained in them—and as we saw in the previous chapter, the role that a molecule plays in a living system does not depend on where it came from—if a molecule is going to cause health problems, people should not be exposed to it, no matter where it comes from. After all, we routinely test our food supply for all sorts of harmful substances, from chemicals to bacteria, for this very reason. Given the current screening techniques for GMOs, I find it hard to believe that hitherto unknown risks are more likely to come from them than from a new agricultural hybrid. In my opinion, good chemical inspection programs protect people, not pieces of paper.

The near-term future for the kind of genetic modification already in widespread use in the United States is pretty clear. Agricultural experts estimate that there is about a 10 percent cost advantage to using these crops today, and they will surely continue to be widely used in the Western Hemisphere. Some wealthy countries, such as those in Europe, may decide that they can afford to pay the extra cost of using crops developed in the old way, just as some people decide that they will pay a premium for organic foods. That is certainly their privilege.

Third World countries, however, cannot afford this luxury. There, an increase in food costs results in very real human suffering and even starvation. The point of genetic engineering is that, while it may be a technology that is difficult and expensive to develop, it is very easy to exploit thereafter. Farmers grow GMOs the same way they grow everything else. The difference, of course, is that over time the GMOs produce more food. In a recent study in India, for example, small farms planted with Bt cotton saw productivity rise a full 80 percent over nearby farms using conventional seeds. In China and the United States, the increase in yield is more modest—typically 10 percent or so—but the decreased need for pesticides raised farm income considerably. And while such increases may not mean much in rich countries, in poor ones they can represent the difference between being able to feed a family and starvation.

The real advance in genetic modification, however, will not come from conventional agriculture, but from a second wave of plants already being developed. One example of this new wave is what are called neutraceuticals. These are food plants that have been engineered to produce molecules that are specifically beneficial to humans. You can imagine, for example, a banana whose DNA has been modified so that it that produces the recommended daily allowance of vitamins. Once such trees are planted, they would continue to produce the vitamins without any further intervention—there would have to be no food-processing factories (and added costs) to enrich the diet of local people. In fact, I can remember a seminar on this subject in which the speaker talked about engineering the bananas to be different colors—"Eat one red banana daily for vitamins, a blue one once a month for protection from cholera," and so on. The beauty of genetic engineering is that once the red banana has been produced, it costs no more to grow than any other kind. It's hard to imagine anything that would have a more positive effect on Third World health than developments like these.

In fact, a strain of genetically engineered rice—called Golden Rice—has already been developed by mixing two genes from daffodils and one from the bacterium *Erwina uredovora* with rice DNA. The rice produces beta-carotene, and its use is expected to decrease blindness and malnutrition associated with vitamin A deficiency. The strain has also been crossed (by conventional means) with a strain of rice that produces iron so that both benefits can be derived from the same crop. Whether the benefits from Golden Rice will be realized by widespread use is still an open question as of this writing, but if it is eventually used on a large scale, it will be a good example of the improvement of agricultural products for the benefit of people in developing countries.

Chris Field, director of the Department of Global Ecology of the Carnegie Institution of Washington, points out ways that genetically engineered plants could also be used in the task of ecosystem management. "In the future," he says, "we may well have 'reporter genes' built into certain plants—genes that will turn on when certain environmental conditions are met." A plant, for example, may be engineered to fluoresce when phosphorus levels in the soil reach a certain level. Monitoring the plants from a satellite would then give us real-time readouts of widely scattered and inaccessible ecosystems.

More importantly, he points out, we can engineer plants to help us correct ecosystem problems. For example, human beings have come to dominate the nitrogen cycle of our planet, and some ecosystems are becoming overloaded with nitrogen. Right now, plants respond to an excess of nitrogen by cutting back on root growth—after all, why bother growing the roots if people will deliver the nitrogen to your door? It would, however, be relatively easy to engineer this trait out of some plants, so that they would continue to remove nitrogen from the environment even when there is an excess supply.

Once you start thinking this way, there is no limit to what can be imagined. You can imagine plants engineered to produce the fatty acids contained in fish meal—fatty acids that are essential in all sorts of animal feed and which now come from the diminishing stocks of fish taken from the ocean. Some scientists have even suggested that you could engineer plants to produce petroleum—a development that would have interesting consequences for the world's energy picture. (Note that carrying these kinds of genes would put a plant at a disadvantage in a world dominated by natural selection, so we wouldn't have to worry about the genes spreading.)

But the most imaginative idea that I have ever encountered in this area actually dates from a time before genetic engineering was a reality. Physicist Freeman Dyson, easily one of the most inventive minds of the twentieth century, wrote an essay in which he asked what it would take to have trees growing on the surface of asteroids. There's nothing less plant friendly than an asteroid's surface—the hard vacuum of deep space, temperatures lower than anything on Earth. On the other hand, there is plenty of energy in the form of sunlight. Dyson began by thinking about how a tree would have to be modified to survive in this environment (as I recall, the trees were supposed to be growing on the outside of asteroids that had been hollowed out for human habitation). Clearly, the leaves of these trees would have to be modified to prevent them from losing moisture to space. I thought of the leaves as being armor plated. The roots would have to be modified so that carbon dioxide would be brought in from the hollowed-out interior of the asteroid and oxygen released the same way. In the end, he imagined a reengineered tree that could, indeed, survive in deep space.

Could we engineer such a tree today? Of course not. Will we be able to do it in the future? I wouldn't bet against it. Once we get really comfortable with modifying living things to meet our needs, this task (and countless others) will be within our reach.

Playing God

By now, I suspect many readers are starting to feel a sense of unease about what I have been saying. It is easy to get carried away with technological promise, but the closer we get to the basic mechanisms of life, the more a sense of "maybe we shouldn't go here" begins to develop. I'm sure that my own optimistic (some would say triumphalist) attitude causes some of my readers to worry about human beings "playing God" by interfering with the natural order of things.

It's easy to make facile comments about this feeling—after all, the development of agriculture was a kind of playing God, as was the development of antibiotics. I believe, however, that this kind of dismissive response (common among my fellow scientists) ignores a very real and very human response to the awesome responsibilities that go with our new knowledge. In a sense, we really do seem to be transgressing into the realm of God or, in alternative language, into the realm of the spiritual.

But while I can sympathize with this unease, I really think our current situation demands that we move beyond it. To me, the entire history of our race consists of moving things from the unknown realm of the gods to a place where they can be understood and used to improve the lot of humans. In many instances, this process is accompanied by a sense of loss, a sense of transgressing boundaries that shouldn't be crossed. For example, lightning—dramatic, unpredictable, deadly—used to be thought of as a tool of the gods (or of a particular God). Benjamin Franklin showed that, instead, it is just the flow of electrical current, and he used that knowledge to invent the lightning rod, a device that has saved countless human lives. At the time, many clergymen opposed the use of the lightning rod as being sacrilegious, an attempt to thwart the will of God. In effect, they accused Franklin of playing God. Today, of course, we don't even notice lightning rods on buildings and power lines.

I suggest that our descendants will someday be as comfortable with genetic technology as we are with lightning rods. Once we get used to it, we will see gene manipulation as just one more step in the process by which humans improve their condition by utilizing our knowledge of how the world works.

12

The Emergence of Complexity

Whenever scientists acquire a new mathematical tool, previously inexplicable parts of the universe become accessible. In the seventeenth century, the branch of mathematics known as calculus was developed by Isaac Newton and the German philosopher Gottfried Leibniz. This mathematical tool allowed scientists to write down (and, sometimes, solve) the equations that governed systems in continuous change. It was, in fact, one of the basic ingredients of the scientific revolution. (There is a long-standing controversy, fueled in part by Newton's litigious nature, about which man deserves recognition for being first to make this discovery. For our purposes, we can consider it a discovery made independently by two brilliant mathematicians.)

Here's an example to show why new mathematical tools make a difference: if you are driving in your car at a steady speed—fifty miles per hour, for example—and want to know how far you will travel in a given time— let's say thirty minutes—you know how to proceed. You multiply the speed (fifty mph) by the time (half an hour) to get the answer. In this case, because the speed of the car doesn't change, you know you will be twenty-five miles down the road when the time is up. In this case a simple mathematical tool—the process of multiplication in arithmetic—gets you the answer you need.

But what if the speed isn't constant? What if the car is slowly accelerating, so the speedometer needle never sits still? How can you predict the distance you will go then? The answer: without calculus you can't. The mathematical tools of simple arithmetic simply aren't up to the job of tackling this more difficult problem. The advent of calculus, however, provided the tools that enable you to solve it, as countless first-year physics students have learned to their sorrow. In fact, calculus and its successors allowed scientists to deal with systems like the sun and its planets (where the position of each planet is always changing), the flow of heat in an engine (where energy fluxes and temperatures are constantly changing), and electrical power lines (where current is constantly changing). Our ability to guide a space probe to a distant planet, to take a CAT scan, or to have a light go on when we flick a switch all depend, at bottom, on the development of calculus centuries ago. In fact, it's hard to imagine modern society without it.

But for all the realms of nature that calculus opened, there were some areas that remained a closed book. In these regions, ordinary calculus was an insufficient tool, just as simple multiplication was insufficient to describe an accelerating car. Imagine, for instance, stretching a rubber band. Exert a small force and the rubber band stretches a certain amount; exert twice as much force, and the rubber band stretches twice as much. In this situation, we say that the rubber band is a linear system, and simple mathematical tools like algebra and calculus will tell you how hard you have to pull to stretch the band a certain amount, how much energy will be sorted in the stretched band, and so on.

Suppose, however, that you stretch the rubber band too far? In this case, the response is quite different. The band becomes flaccid, unable to resist further stretching. A small force will result in a disproportionate lengthening of the band—indeed, may even break it. In ordinary language we say we have overstretched the rubber band, or broken it. In technical language, we say that we have exceeded the "elastic limit." This sort of two-tiered response is said to be nonlinear, and there are may examples of nonlinear systems in the world. If you turn up the volume on your radio or CD player, the response at first is linear—turn the knob twice as far and the music becomes twice as loud. In this range, the response of the system is nicely linear. Keep turning, though, and you start to hear distortions and squawks. At this point, the response has become nonlinear, with changes in output being disproportionate to changes in the controls.

For technical reasons, nonlinear systems are extremely difficult to handle with the tools of calculus, just as the accelerating car is difficult to discuss with the tools of arithmetic. In a few special cases, nonlinear systems (and the nonlinear equations that describe them) can be solved by special mathematical tricks, but for the most part, scientists from the time of Newton to the mid-twentieth century either focused their attention on linear systems (which they could describe with the tools they had) or had to be content with various approximate methods to deal with nonlinearity. Like the old medieval maps, the nonlinear domain was largely a blank, perhaps adorned by an ornamental "Here be Monsters."

This unknown area included the science of ecosystems. There are, in fact, many examples in which a small change (the introduction of rabbits in Australia, for example, or the Dutch elm disease fungus in North America) led to a huge effect (the denuding of rangeland or the decimation of forests). If our goal is to manage the planet and its ecosystems, we are obviously going to have to predict how nonlinear systems respond to changed circumstances, and this means that we have to develop a new set of mathematical tools, just as Isaac Newton and his contemporaries did so long ago.

The development of calculus was a pencil-and-paper operation. Scientists wrote down the equations that described the system they were investigating, then found ways to solve those equations in terms of standard mathematical functions—things like the sines and cosines you may remember from your trigonometry course. Numbers came into the picture only at the end, when you already had your answer and began looking up the values of your mathematical functions in specialized tables. (Did you know that the word "computer" used to refer to people whose job it was to look up and manipulate those numbers?) The problem with nonlinear systems is that this tried-and-true method doesn't work, because the equations that describe them can't normally be solved in terms of the familiar mathematical functions.

Enter the digital computer.

The way a computer (the machine, not the human being) goes about dealing with a nonlinear system is different from the Newtonian pencil-

and-paper operation, just as the way that a computer plays chess is different from the way that a human plays. The computer puts numbers in right at the start—after all, numbers are what computers are good at. It starts with the numbers that describe the system at a given point—it might be a description of temperatures, winds, and barometric pressures at noon on a given day, for example. It then uses the equations that describe the system to step forward a small interval—a few minutes in the case of a weather system. It now knows the numbers that describe the system a few minutes after the initial time, and it uses these as a base to step forward a few more minutes in time. Eventually, using the blinding computational speed available to it, it comes up with a statement about the numbers that describe the system at some time in the future. In this example, those numbers might be a prediction about the temperatures, wind speeds, and barometric pressures three days from now.

Could human beings, working with pencil and paper, do the same calculations? Of course. The problem is that it could well take them years to predict tomorrow's weather, and the lifetime of the universe to predict the climate a century from now. This is why the exploration of the nonlinear world had to wait until the advent of the high-speed computer.

And, as always happens when a new domain of the universe is opened to scientists, there were a lot of surprises waiting for us. The two most important surprises were the discovery of the phenomenon of chaos and the beginning of an understanding of complex systems and the related phenomenon of emergence.

The Chaotic World

One of the features of the Newtonian view of the world goes by the philosophical name of determinism. As the term is used by physicists, a system is deterministic if, given its state at one moment of time, you can predict its state at any time in the future. From the discussion above, it is clear that both linear and nonlinear systems can be deterministic.

One of the first great surprises that came out of the exploration of the nonlinear world was that there are some systems that, while deterministic in principle, are, in fact, not deterministic in practice. Such systems are called chaotic, and a simple example will help you understand how such systems work.

Think about a swiftly flowing river coming in to an area of rapids, with white water everywhere. If you put something in the water on the upstream side—a little piece of bark, for example—it will churn its way through the rapids and come out somewhere in the calm water on the downstream side. If you wanted to take the time and the energy, you could probably trace out the path that the bark would take and predict where its exit point would be.

Now suppose that after this exercise you put another piece of bark down near where the first piece had started its journey. If you put the second piece down in *exactly* the same place as you had put the first, it would trace the same path through the rapids and come out in *exactly* the same place as the first. If, however, your placement is off by the smallest amount—a gnat's eyebrow—the path through the rapids will be different, and the second piece will come out somewhere far from where the first piece did. A system like this, in which the outcome is exquisitely dependent on the details of the initial conditions, is said to be chaotic. (Technically, a chaotic system is one in which the final state of the system changes exponentially with small changes in the initial state.)

In every measurement, such as the measurement of the position of that first piece of bark on the water, there is some uncertainty. If nothing else introduces it, then the fact that you are measuring the position with a ruler indicates that you will be uncertain about the bark's position at a level below the ruler's smallest division. For example, if your ruler is marked off with lines a sixteenth of an inch apart, you will not be able to describe the position of the bark to within a thousandth of an inch. But the point of a chaotic system is that if two pieces of bark start off a thousandth of an inch apart, they will wind up at different places on the downstream side of the rapids. Thus, for all practical purposes, we cannot predict the future of a chaotic system.

Let me say this in another way, just to make it clear. It is possible to predict the future of a chaotic system if you know the initial state of the system exactly. In the real world, however, it is never possible to have exact knowledge of that initial state, so for all practical purposes, the future cannot be predicted. Physicists have taken to referring to this phenomenon as deterministic chaos, to emphasize both the predictability in principle and the unpredictability in practice.

The story of the discovery of chaos illustrates the sensitivity that these systems can have to small changes in the initial conditions. In the 1960s, Edward Lorenz and his colleagues at MIT were running computer models of the Earth's climate (simple forerunners of the GCMs discussed in chapter 9). Sometimes they could get the computers to do the calculations in a single long run; other times they would run for a while, store intermediate results on perforated paper tape, and then feed that tape into the computer later to finish the calculation. (In those days, computers were rare, and you often had to interrupt calculations to let someone else use the machine. I can remember, as a summer student assistant at Argonne National Laboratory near Chicago in the 1960s, coming to work at 3 A.M. so I could use the computer.)

In any case, the scientists noted that the predictions that came from the computer in these two situations were wildly different from each other. This was strange—why should two ways of keeping track of the numbers produce different predictions for the weather? Eventually, they traced the difference to the fact that the computer rounded off numbers differently when it was transferring them to tape and when it was using them in an ongoing calculation. The difference was tiny—a change in the fourth decimal place—but was enough to change the predicted outcome. This finding was the first intimation scientists had that such a thing as chaos existed, that there were systems where changing the fourth decimal place halfway through a calculation could produce a radically different result for the final outcome.

Having made such a big point about how small changes in input can produce huge changes in the future state of a system, I need to bring the discussion back to Earth. Chaotic systems may be unpredictable, but they're not *that* unpredictable. Think again about the piece of bark on the rapids to see what I mean. If you start two pieces off right next to each other, it is true that they will come out far apart on the downstream side. It is also true, however, that they will stay pretty close to each other for a while. So even though it's true that eventually the two pieces of bark will move away from each other, it's also true that for a while they will stay close together. This is also a general feature of chaotic systems.

What this means is that for a given uncertainty in the measurements that determine the initial state of a system, there will be a period of time during which the eventual chaotic behavior of the system can be ignored, a period during which we can get an approximate prediction of its future state in the old-fashioned Newtonian sense. Provided that we are content with solutions that are good enough, rather than mathematically precise, we can still predict the future behavior of even these systems.

For example, there is an ongoing debate among scientists about whether or not the Earth's weather and climate systems are chaotic in the mathematical sense or simply very complicated. Even if we assume they are chaotic, however, we can still make pretty accurate predictions of the weather for a period of days into the future. The reason, of course, is that, like those pieces of bark on the stream, the predictions for future weather for slightly different initial states stay together long enough for us to make good enough predictions. There are other places where this same phenomenon can be seen. For example, astronomers calculating the behavior of the solar system, including all the thousands of asteroids, moons, and other objects as well as the planets, have concluded that it is a chaotic system, but that predictions of future states diverge on a timescale of *billions* of years. In this case, the "pieces of bark" stay comfortably near each other for a long time indeed, which is why astronomers can predict the dates of eclipses thousands of years in the future.

This property of chaotic systems is also extremely important for the future of ecosystem management, because it tells us that we don't need to have perfect knowledge of any system to make good enough predictions of what the short-term effects of a particular policy choice will be. We just have to build in the ability to make midcourse corrections before time runs out on our predictive abilities—before the two pieces of bark get too far apart.

Probably the most widely known folklore about chaotic systems has come to be called the butterfly effect. It goes like this: the Earth's weather system is so chaotic, so finely balanced, that a butterfly flapping its wings in São Paulo can eventually cause a thunderstorm in Moscow (you can substitute your own pair of cities here).

This is one of those statements that is both (1) completely true and (2) totally misleading. It is true because if you imagine two identical worlds, one with the flapping butterfly and one without, their weather patterns will eventually be different from each other, just as the pieces of bark move apart in the rapids.

On the other hand, that flapping butterfly will cause the storm *only if there are no other butterflies flapping their wings anywhere on the planet.* In the real world, the effects of that one butterfly get averaged in with the effects of all those other butterflies, along with everything else that affects the weather. This explains why the Earth's weather, unlike the water in the rapids, tends to follow a fairly stable pattern, and why, in the end, weather prediction is possible.

Complex Systems

Along with chaos theory, the emerging field of complexity theory has transformed the way we understand the world. In mathematical terms, a system is said to be complex if it contains many agents, each of which influences and is influenced by the others. A stock market is the classic example of a complex system, because each player's decision to buy or sell influences the actions of other players, whose decisions, in turn, influence the original person. Ecosystems are also complex in this sense, since what happens to one species can affect others, and that effect will come back to influence the original species. Perhaps the most complex of the complex systems we know about is the human brain, where each neuron is connected to up to a thousand other neurons. Each neuron integrates the impulses it receives from all of its neighbors in order to decide whether to send out impulses of its own, and that decision, in turn, influences all of the neurons to which it is connected.

In the mathematical sense, the words "complex" and "complicated" are not synonyms. There are many complex systems that are pretty simple, and many complicated systems that are not complex. A simple pile of sand grains, for example, is a complex system because it contains many grains (agents) that influence each other (by pushing on their neighbors). On the other hand, you can build very complicated electrical circuits that are not complex because different parts of the circuit don't respond to the actions of other parts.

Computers are necessary for understanding complex systems because they are good at keeping track of many things—think of the computers that run an airline ticketing system, for example, or the orders for a major retail firm. Thus, like the study of chaos, the science of complexity is a fairly new entry into the mathematical arsenal of the human race. Consequently, we don't know exactly how everything will turn out—what this new tool will allow us to do and what it won't. Nevertheless, we already know enough to see some of the major features of this new landscape.

Take the example of the sandpile as a starting point. In the pile, each sand grain pushes against its neighbors, and each is, in turn, pushed on. Think about what happens as you add grains to the pile. For a while, the extra grains just add to the overall weight, with perhaps some minor shifting and jostling as grains already present adjust to the presence of the newcomer. There will, however, be a point at which adding one new grain produces an entirely new kind of behavior—a landslide. This new behavior is called an emergent property of the sandpile and is an example of the most interesting aspect of complex systems.

Emergent properties show that there are systems where "more" eventually becomes "different." The old saying about the "straw that broke the camel's back" is a recognition of an emergent property in folklore, since a small addition (a piece of straw) changes the state of the camel in a fundamental way.

The appearance of an emergent property is not a simple linear extrapolation of the behavior before that emergence. If the landslide starts when you add the millionth sand grain, for example, it is simply not true that you can get a millionth of an avalanche from a single grain. Emergence is a true property of complexity—you see it only when the system reaches a certain size or richness of interaction. Many seemingly inexplicable phenomena—stock market panics, the development of consciousness in the brain, crashes of populations in ecosystems—have been suggested as emergent properties of the systems of which they are part.

The most important question about emergent properties is simple: if we know the properties of the agents in the system, can we predict emergence? Can we, in other words, start with properties of sand grains and predict the onset and characteristics of the avalanche to come?

The short answer to this question is not yet, although in some cases (such as the sandpile), we can make some kinds of predictions about the avalanches—we can, for example, make statements about the statistical distribution of the sizes of avalanches in a large number of sandpiles. In general, though, predicting the onset and character of emergent properties remains a task (some would say *the* task) for the future. And, to be fair, we should point out that until we have a full-blown theory of complex systems, we can't be absolutely sure that making such predictions is possible.

Actually, this discussion illustrates something very important about the future of science. I like to think of scientific knowledge as being arranged on a series of floors in a tall building—think of a department store with men's clothing on one floor, women's on another, shoes on yet another, and so on. The floors of the scientific building have labels like "atoms," "materials," "cells," "organisms," "ecosystems," and so on. The history of science for the past several centuries has basically involved exploring each of these floors more or less in isolation from the others. We've done a good job of this. We really do understand a lot about what is on each floor, and what we don't know now, we're learning fast. In all of this enterprise, however, we haven't paid much attention to the staircases, the elevators, or the escalators—the links from one floor to the next.

In fact, when I think about the sciences, I see only one functioning "staircase"—the one between atoms and molecules, on one hand, and the properties of materials made from them, on the other. A branch of science called statistical mechanics, developed in the late nineteenth and early twentieth centuries, illustrated how connections between the different floors in our analogy can be made. Knowing something about atoms (their average velocity in a material, for example), this science allows you to say something about an "upper floor" property (temperature).

My sense is that the connection between agents in a complex system and the characteristics of the emergent properties of that system represent another such stairway that needs to be investigated. Whether it will be a clean connection, like the one provided by statistical mechanics, or something more vague and less predictive remains to be seen, of course, as does the question of whether such a stairway exists at all.

Completing work on this stairway may be important for the future of the managed planet, because our ability to understand how a complex ecosystem will respond to a perturbation might well depend on knowing when we have to be careful about triggering emergent phenomena—the sort of changes in the role of the Great Conveyor Belt in possible future climates (see chapter 9) might be an example of this. It could be, for instance, that small changes in things like temperature or rainfall or the presence of nearby developments would have relatively minor effects on a given ecosystem, but that when these perturbations reach a critical point, something new and unexpected will happen. Knowing where that critical point is, even if we don't know in detail what will happen, would be a useful piece of information for the planning process.

However, it is important to realize that as useful as a full-blown theory of complexity and emergent properties could be, it's not really essential for managing the planet. As we saw in our discussion of chaos, if an approximate understanding of the system is good enough, we don't need the full prediction. In the same way, a partial theory of complexity that allowed us to predict the onset of an emergent phenomenon even if we couldn't describe the phenomenon in detail would probably be good enough for management purposes. It might tell us, for example, that we could log a certain portion of a forest without causing harm, but that the ecosystem would crash if we went past a certain point. And, of course, any theoretical understanding of emergence would be bolstered by accumulated experimental and observational information about ecosystems, the subject of the next chapter.

13

Experimental Ecosystems

There are basically two ways of tackling difficult scientific problems: tinkering and theorizing. Of these, the first is the oldest method used in the sciences, while the second has come to dominate the thinking of scientists in the latter half of the twentieth century.

When confronted with something like an ecosystem, which may be quite complex and difficult to understand, someone following the tinkering approach will poke and prod, trying to get a feel for how things work. Little by little, this kind of knowledge accumulates until we have a pretty good idea of how the system will respond in different situations. Initially, the new information will be cast as a set of empirical rules, with no accompanying deep understanding of why the system behaves as it does. That understanding may or may not come later.

A theorizing approach, on the other hand, begins with the basic laws that govern physical systems—laws of chemistry and physics—and tries to build upward to an understanding of the system. The difficulty with this approach, of course, is that the more complex a system is, the more difficult it is to see how the basic principles are to be applied.

Historically, most areas of science have evolved from tinkering to theorizing. As we have gotten better at dealing with mathematical modeling, however, and particularly since we have gotten access to fast computers, this has started to change. As we saw in the area of climate modeling, many fields in modern science now begin the analysis of complex systems by writing down the basic equations that describe parts of the system, and then using computing power to put all the pieces together.

There is, in fact, a good historical example of this progression from tinkering to theory. In the late nineteenth and early twentieth centuries, it became very important for scientists and engineers to understand the way that fluids and gases behaved when they moved past solid objects—think of the flow of air over the wing of a plane as an example of this process. As a result, two separate fields of science were developed. On the one hand, the basic equations that govern the flow of fluids were well known, but could be solved mathematically only for fluids whose motion was unimpeded by solids and for a few simple cases where solids were present (remember, there were no computers in these days). The field devoted to the study of these situations was known as hydrodynamics, and hydrodynamicists worked on problems like explaining ocean waves and the flow of the air around the planet. This field was an example of the theoretical approach to the problem.

On the other hand, practical engineers, charged with designing machines that would actually work when real fluids met real solids, began developing a kind of rough, rule-of-thumb set of rules that described how specific fluids behaved under specific conditions—how a lubricating oil would be squeezed out between gears, for example, or how water would flow around the hull of a ship. Known as hydraulics, this field represented a tinkering, practice-driven approach to fluid flow. It wasn't based on general principles, but it worked.

For most of the twentieth century, hydrodynamics and hydraulics remained distinct, and, as far as I can remember from my student days, people in one field rarely talked to people in the other. The availability of high-speed computers changed all this, however. The mathematics that made it difficult for scientists to work out the flow of air over an airplane wing, for example, became a simple calculation once they could apply the power of modern computers to the job. Because of this development, the two fields began to merge, and today they are virtually indistinguishable.

Perhaps the best illustration of this change can be seen in the demise of the wind tunnel. This was a device that generated high-speed airflows. Models of aircraft were put into the tunnel to see whether they behaved as the designers wished. This approach, of course, is a classic example of tinkering. But starting with the 747 in the 1970s, aircraft no longer have been subjected to wind tunnel tests. They are now designed by computers that calculate the way the aircraft will behave when subjected to real flight conditions, based on our accumulated data on how air and airplanes interact. Given the safety record of airlines over the last few decades, this theoretical approach seems to be working quite well.

The same sort of thing has been going on in all sorts of fields, to the point where some scholars have suggested that computer modeling of physical systems should be considered a separate, third branch of science, complementing the traditional fields of experiment and theory. So successful has this sort of work become that, for a while in the 1980s, I began to fear that scientists were losing the ability to tinker and were being completely seduced by the power of the computer.

But successful modes of operation never really die. Faced with the new challenge of understanding complex ecosystems, many scientists have reverted to the old tried-and-true approach of tinkering. There are many groups doing this sort of work, but let me tell you about the study run by ecologist David Tilman, Distinguished McKnight Professor at the University of Minnesota, which has as its goal an understanding of the ecosystem of the midwestern prairie.

Tilman is a tall, soft-spoken man who has achieved a high standing in the still-new study of ecosystems and is often called on for advice by various government and private agencies. His office is in a modern building, but it isn't located at the familiar Minneapolis branch of his university. Instead, it is at the St. Paul campus, which houses the agricultural school and the university's experimental farms. Having been an undergraduate at the University of Illinois, which has an experimental corn plot in the middle of its campus, I felt right at home when I visited this place a few years ago. It was a bright spring day, and the place was overrun with clean-cut kids wearing Future Farmers of America jackets—I had apparently invaded their annual meeting.

Tilman began the study of the prairie ecosystem in a large, open field located at the Cedar Creek Natural History Area, about thirty miles north of Minneapolis. He divided the field into 207 plots about twelve feet on a side and began to experiment. In one of his early works, which has become, literally, a textbook example of how to study ecosystems, he studied the effects of adding different amounts of nitrogen to the plots.

Why nitrogen? The growth and well-being of every organism on the planet is limited by something. Humans, for example, need to take in a certain number of calories every day to continue to function. We also need to take in a certain amount of various vitamins and minerals—if they are missing, the results can be serious. Think, for example, of a disease like scurvy, caused by the absence of vitamin C, or various kinds of blindness caused by deficiencies in vitamin A.

In just the same way, plant growth is always limited by the availability of some resource in the environment, whether carbon dioxide, water, or some essential element. For many plants, limitations of growth and well-being are set by the availability of nitrogen. This may seem strange, because nitrogen is one of the most common chemical elements around—80 percent of the air you are breathing in right now is nitrogen, for example. The problem is that individual nitrogen atoms normally form very strong chemical bonds with other nitrogen atoms—bonds so strong that most living things can't pry them apart. There are both natural and industrial ways of separating these atoms—of "fixing" the nitrogen, in the jargon of chemistry—so that plants can use them. When you fertilize your lawn or garden, you are supplying the plants with fixed nitrogen to help them flourish.

Tilman decided to find out what would happen to ecosystems that were supplied with progressively more nitrogen in fixed form. He left some of his plots alone—in effect, let them get by on whatever nitrogen nature would supply them. Other plots were supplied with various nutrients like phosphorus and potassium, but no nitrogen. To still other plots, he added nitrogen in different quantities, from small amounts to virtual saturation, and then observed what happened. This, of course, is the classical scientific method, applied to a complex system.

What he found will not come as a surprise to any experienced gardener. As the amount of available nitrogen increased, so, too, did the amount of vegetation—what scientists call the biomass—in each plot. This is reasonable. If plant growth is limited by nitrogen availability, then making nitrogen available will increase plant growth.

He also found that as the biomass in each plot increased, the number of different species of plants—what scientists call the biodiversity—began to drop. Again, this is a reasonable result. Some plants have evolved to get along on small amounts of nitrogen and are incapable of absorbing more. To them, the extra nitrogen is simply irrelevant. Other plants, however, are limited by nitrogen and respond to increased availability by growing quickly. When competing against each other, the latter will crowd out the former in a fertilized plot.

In 1987 and 1988, a fortuitous event shed further light on the prairie ecosystem. Central Minnesota went through a severe drought, so suddenly the plants in all the plots were limited primarily by the availability of water, rather than by nitrogen. In this case, adding nitrogen had only a small effect on biomass, since it was no longer the limiting factor. Biomass dropped precipitously for all plots (again, as expected), but the drop was proportionately larger in the plots that had been highly fertilized (again, no surprise, since the biomass was falling from a higher value). Thus, it appears that biodiversity provides a kind of insurance against disaster—with more species, the chance increases that some will make it through hard times.

From these sorts of studies, we can begin to get a sense of how this particular ecosystem works, which can lead to insights into how it might be managed. If the goal is to produce biomass (as it might be for a farm field), then it should be fertilized (and watered if necessary). If the goal is to have the system survive on its own, without human intervention, then you want to have high diversity.

By drawing on work like Tilman's and that of other experimental ecologists, work that is firmly in what I have called the tinkering tradition, we will begin to get a handle on the workings of even the most complex ecosystems. In the end, this heuristic knowledge will be translated into mathematical models and theories, but right now the field is very much engaged in this data-gathering, trial-and-error phase. When theories are developed, however, a crucial piece in our ability to manage the planet will be in place.

The Law of Unintended Consequences

The most common reaction to such a statement is disbelief. Skeptics argue that ecosystems are just too complicated to be managed, that no matter how careful you are, you will overlook something that will destroy your planning. This argument is often couched in terms of something called the law of unintended consequences. This law is actually a variation on Murphy's Law ("If something can go wrong, it will") applied to ecosystems. (There actually was a Murphy. Captain Edward Murphy, an engineer, worked on early experiments to test human reactions to rocket acceleration. In what has to be one of the most striking examples of Murphy's Law, his original statement is always misquoted. His actual words were "If there is more than one way to do something and one of those ways won't work, someone will come along and try it.")

There are certainly many examples of this law in action. Rabbits brought to Australia as a food source for transported prisoners in the eighteenth century, for example, blossomed into one of the greatest agricultural problems on the planet. The importation of kudzu into the southern United States to plant on road cuts quickly loosed a stubborn nuisance on an unsuspecting nation. Even apparently innocent actions can have unintended consequences. During the nineteenth century, for example, there was a group in the eastern United States devoted to bringing every plant and animal mentioned in Shakespeare to North America. I guess it seemed like a good idea at the time, but one day they turned loose a group of starlings in Central Park in New York City. The birds multiplied and are now one of the main (and often unwelcome) components of urban ecosystems in the United States.

The textbook example of unintended consequences involved the introduction of a fish known as the Nile perch into Lake Victoria in Africa. The idea was to introduce a species that would support sport fishing. At the time of its introduction, the staple catch of fishermen along the lake were a set of smaller fish species native to the lake. These fish were dried on racks in the sun, and producing these dried fish was the main economic activity in the area. When the Nile perch were introduced, they fed on the smaller fish, quickly reducing their numbers. The fishermen then turned to the perch, but they were too big to be dried in the sun—they had to be roasted. This, in turn, required that the fishermen find a source of firewood, which they did by cutting trees and woody plants growing on the shores of the lake. As a result, the land, deprived of its cover, began to erode. This progression—from introducing a species for sport fishing to eroding the land—captures the essence of what people mean by the law of unintended consequences.

The argument, then, is that no matter how carefully you plan, no matter how complex your model is, there is no way to anticipate everything, and something like the Nile perch story is bound to happen. Because of this, the argument goes, we should leave nature alone and not interfere. This is the "don't rock the boat" approach that I mentioned in my discussion of pop ecology.

How does one respond to this sort of argument? At one level, I could point out that many of these examples date from a period before people really thought very much about the environment. They tend to represent a kind of thoughtlessness, rather than being examples of some sort of inherent human inability to manage complex systems such as the Earth. We do a lot better when we pay attention to what we're doing.

At another level, I can point out that the history of any large engineering process always begins with mistakes. In fact, from the point of view of the engineer, you *want* to see failures at the beginning—how else are you supposed to find and fix the weak points in the design? Take the U.S. space program of the 1960s as an example. We tend to remember this program as a steady triumphal march from President Kennedy's speech setting an American moon landing as a national goal to the 1969 landing itself. In fact, the early days of the American space program were dogged by one failure after another. While the Soviet Union put satellite after satellite into space, American rockets were blowing up on the launchpad.

This discouraging period in our nation's history is recounted in detail in Tom Wolfe's marvelous book *The Right Stuff*. Anyone observing the morose sequence of events might well have concluded that building a successful rocket was simply beyond American capability, that something like the law of unintended consequences would always keep us from our goal. But this attitude overlooks the greatest strength of the engineering mentality. Each failure highlights a design flaw, which the engineer seeks out and fixes. The next failure highlights another flaw, which is treated similarly. Eventually, you run out of flaws and the system starts to perform its function as it should.

As an example of this process, consider the sequence of events that led up to the Soviet Union's successful launch of *Sputnik,* the first artificial satellite, in 1957:

May 15—	the R7 rocket explodes one hundred seconds into flight (ruptured fuel line)
June 9—	rocket fails to fire (valve installed upside down)
July 11—	rocket goes off course and has to be destroyed (faulty guidance system)
August 21—	first successful launch (rocket travels four thousand miles)
Oct 4—	*Sputnik* put into orbit

Knowing that things work this way often gives engineers a somewhat gloomy mind-set—a fact that was brought home to me a number of years ago, when I was on the faculty of the University of Virginia. The man in charge of designing the engines for the space shuttle (which was not yet a reality) came to give a talk. For an hour and a half, he kept an auditorium full of engineers enthralled with a list of all the reasons why the engine could never work. I want to tell you, those guys were in hog heaven—all those problems! Of course, within a few years, the engine worked perfectly, as everyone knew it would.

In the same way, I suspect that the horror stories about unintended consequences are the ecological analogue of those exploding rockets. We've made mistakes and learned from them, and we will undoubtedly make mistakes in the future. We'll learn from them, too. And, like those exploding rockets, our mistakes in ecological management will carry a price: that's part of the way things work. For the record, I think we're getting pretty close to the end of this process in terms of ecosystem management. Consequently, I really don't worry too much about the law of unintended consequences.

There is one more point I'd like to make on this subject. In my experience, the law of unintended consequences is most often encountered when a group or an individual wants to prevent human intervention in some natural ecosystem. "Better let nature take care of itself," the argument goes, "because if we humans try to do something, we'll just screw it up."

But deciding not to intervene is itself a decision made by humans, and it, too, may have unintended consequences. We can't get out of the responsibility to manage the planet by hiding behind the law—if it operates, it operates no matter what you do. Damming a river may have unintended consequences, but so may not damming it. Better, I think, to create a science that can reduce unintended consequences than to hide behind the "law" as an excuse for doing nothing.

The question of whether or not we will be able to manage ecosystems depends, in the end, on whether we are able to predict what effects a particular course of action will have. And this lands us squarely in the middle of an old debate in the sciences, which we can characterize roughly as a difference in approach between the physical and the life sciences or, to follow the lead of ecologist John Harte and personalize the issue, between Isaac Newton and Charles Darwin.

In a sense, this difference is a matter of focus. When a physicist looks at a problem, he or she immediately starts to pare away inessential details while searching for the fundamental underlying simplicity that is supposed to be there. (The model-building procedure described in chapter 9 is an example of this approach.) This way of looking at questions, which we can trace back at least as far as Isaac Newton, has successfully illuminated a large part of the physical world.

A biologist or ecologist, on the other hand, may look at the same system and focus on the complexities themselves. To him or her, what is interesting about the system is its complexity, its interrelatedness. To a life scientist, a physicist's models are little more than caricatures of reality, at best amusing, at worst misleading.

Another difference in outlook can be characterized as a debate about the relative importance of chance and necessity. When Isaac Newton looked at the universe, he saw something like a clock, with gears ticking along according to known (or at least knowable) laws—a universe whose behavior was completely predictable. When Charles Darwin looked at the natural world, however, he saw a system that had resulted from a series of historical accidents and that was, therefore, essentially unpredictable. In the language of philosophers, he saw a universe dominated by contingency.

The question before us is where between these two extremes ecosystems fall. The question of whether you could predict the existence of the current biosphere—whether, in the late Stephen Jay Gould's words, "If you played the tape again you would get a different tune"—is an interesting one, but not particularly relevant to the problem we are looking at. The question we have to ask is whether or not, given the characteristics of a specific existing ecosystem (a prairie, for example, or a forest), we can predict enough about its response to a perturbation (higher temperatures, for example) to allow us to manage it. This is at the same time a simpler question than that contained in the grand Newtonian-Darwinian dichotomy and a harder one, because the answers we require are more detailed.

Predicting the Planet

So where will the study of ecosystems lead us? Can we eventually develop a science that will allow us to predict the effects of human interventions, or will we always be surprised at unexpected outcomes?

The honest answer to this question is that we don't know. The essence of a scientific frontier is that we don't know what lies beyond it. To make matters worse, when we talk about managing the planet, we are dealing with at least two frontiers—one in ecology and the other in complexity theory. Nevertheless, we can imagine different outcomes and think about their implications for the management of ecosystems. At the two extremes we could be faced with

CASE I: *The behavior of all ecosystems is essentially unpredictable, although it may be understood in retrospect*

CASE II: *The behavior of all ecosystems is essentially predictable*

There are, of course, intermediate possible outcomes between these two extremes. It could turn out, for example, that the effects of changes to some ecosystems are predictable while that of others are not. It could turn out that some aspects of behavior are predictable while others are not, and so on. We will turn to these intermediate—and, to my mind, more probable—outcomes after we have considered these two extreme cases.

CASE I

There is no question that ecosystems are complex systems, as defined in chapter 12. They have many agents (the organisms that make them up), and the behavior of each agent affects the behavior of the others. We expect, then, that ecosystems will exhibit the characteristics we associate with complexity, and, in particular, that they will have emergent properties—sudden changes in behavior—as their complexity increases. To the extent that it remains difficult or impossible to predict the onset and characteristics of emergent properties from the properties of the individual organisms in the system, the behavior of ecosystems will also remain unpredictable.

However, we have to realize that there are other ways that prediction could fail. It could turn out that ecosystems exhibit chaotic behavior, so that even though we could in principle predict how they will respond to a specific perturbation, in practice the amount of precise data we would need to make this prediction is so great that, for all practical purposes, prediction becomes impossible. Or it might simply be that the complexity of the system is so vast that we will never be able to untangle it.

It seems to me that this extreme is unlikely to be the one that corresponds to nature. Ecosystems are complex, but they're not *that* complex. Compared to something like the human brain or the stock market, they are rather simple. In addition, I have seen too many seminars and conferences in which scientists presented sophisticated computer models of ecosystem behavior to believe that prediction is completely impossible. I remember one session at a scientific meeting in San Francisco, for example, when a man from the Department of Agriculture described his predictions for the effects of global warming on forests in the United States. To back up his argument, he nonchalantly showed the results of government computer programs that modeled the nation's forests square mile by square mile. I think I was the only person in the room who was flabbergasted by the fact that such a model existed—everyone else seemed to take it for granted and was concentrating on the details of his calculations. The fact is that while theoreticians and philosophers argue about whether such models can be constructed, people in the trenches are going ahead and building them. Such models exist for forests, the Great Lakes, and most other ecosystems. If ecosystem behavior were really unpredictable, these models wouldn't work. They do work, however, and for my purposes, that's all that matters.

So scientists studying specific systems have been able to tinker their way to general rules and behaviors and, in many cases, have translated that knowledge into predictive models. This is yet another piece of evidence that we can get enough knowledge about ecosystem behavior to manage the planet. In terms of the historical example given above, engineers armed with a knowledge of hydraulics were perfectly capable of designing aircraft that flew safely and efficiently. In the same way, it may be possible to manage ecosystems even if a grand theory of ecology (the equivalent of hydrodynamics) eludes us.

CASE II

While an essentially unpredictable planet is, to my mind, unlikely, I frankly think that a completely predictable one is equally unlikely. The problem, for me, comes from the nonlinear nature of complex systems. The tinkering approach works very well in dealing with small perturbations and small responses, but we also know that complex systems sometimes exhibit large responses due to small but critical changes in initial conditions (think of a stock market crash, for example). Unless our experimental input includes that critical value, we will always have to worry that if we push the system too far, we will move it into the region where a new (and unanticipated) emergent property arises.

There is also the possibility that in some cases our theoretical knowledge may be good enough to understand why a system behaved as it did in the past, but not good enough to predict what it will do in the future. Consider, for example, the history of life on our planet. We can construct a story that tells us how things happened—how the dinosaurs became extinct, how human beings developed, and so on. It is much harder, however, to predict the course of future evolution (at least with our current state of knowledge).

Having said this, however, I have to come back to a point I have made repeatedly in this book. We don't have to have a perfect, well-thought-out theory of ecosystems to manage the planet. Our knowledge doesn't have to be universal, just good enough to do the job. We don't have to be able to predict the entire future outcome associated with introducing a certain chemical into the biosphere. All we need is to know enough to introduce a small amount, see what happens, make any necessary corrections, add some more, and look again.

Because of this, my guess is that the final outcome of the Newtonian-Darwinian debate will wind up bringing us closer to the former than to the latter. There are just too many examples like that Department of Agriculture model of the nation's forests for me to think that good enough working theories must forever lie outside of our reach. To quote Shakespeare's Richard III:

"What, can we do all this and still not win the crown?
Tut, were it farther off, we'd pluck it down"

I suspect that for most ecosystems in most cases, we will be able to predict the effect of what we propose to do, and I suspect that we will also eventually develop a pretty good sense of when we need to be more careful. And in the end, these are the only tools we really need to manage the planet.

IV

The Second Step

14

A Matter of Choices

We are now ready to talk about the momentous changes that are about to occur in the relationship between humans and nature, and how those changes are to be managed. In introducing this topic, I spoke about the first great change—I called it the First Step—that began with the development of agriculture and allowed humans to sidestep natural selection. I called the great adventure upon which we are embarked the Second Step, a step in which humans will reenter the natural world, not as ordinary participants but as managers.

It is now time to turn to what is perhaps the most difficult question we can ask in this area—given that we will have the power to manage the planet, how shall that power be used? I proposed what I called the benefit-to-humans principle, which is that the planet should be managed to maximize the welfare, broadly conceived, of human beings. There are, of course, other possible principles we could use, and it is to that subject that we now turn.

Up to this point in the book, I have tried, as far as is in my power, to present ideas based on science. When I have departed from what I take to be the scientific mainstream (as in the discussion of global warming), I have tried to explain why I have done so in some detail. We have reached the point in the argument, however, where considerations outside of science—values, beliefs, and principles—must enter the picture. We must begin to enter the world of mythos, to go from arguments taken solely from the magisterium of science to arguments that mix together aspects of both science and belief. We must venture, in other words, into the real world in which personal decisions about policy matters are made.

In an English Garden

As a young man, I had the privilege of being selected as a Marshall scholar to study at Oxford. (These are scholarships for American students sponsored by the British government as a "thank you" for the help postwar Britain received from the Marshall plan.) Oxford had quite an impact on my life. After all, here I was, a kid from a blue-collar neighborhood in Chicago who knew, personally, every member of my family who had ever lived in America, suddenly plunked down in a college whose charter was written in 1272. Talk about expanding your time horizon!

On a purely personal level, my experience at Oxford opened my eyes to the existence of a long line of scholars and teachers, extending centuries into the past, and to the possibility that I could be a part of that tradition. One significant experience, however, was more mundane. Merton College, where I was a student, was built along what was once the old city wall of Oxford. Next to the college buildings, which include a marvelous Gothic chapel and library, was a garden. Situated between the old city wall and the outer wall of the college, this was a garden of the type you see only in England. Tall lime trees shaded an expanse of land as flat as the gardener's art could make it, while flowers and shrubs formed borders and paths throughout. There were few things more pleasant at Oxford than sitting in that garden on those long, blue English summer evenings.

But what struck me then about the garden, and what continues to strike me now, is that it has been maintained, in one form or another, for seven hundred years. It represents a totally managed ecosystem that has survived longer than many natural ones. Merton College garden is much older than the hardwood forests of the eastern United States, which only started to replace farmland in the mid-nineteenth century. It was already old when the managed ecosystems in the New World collapsed following the advent of Europeans. Its age, in fact, is comparable to the amount of time the Mississippi River has followed its present channel past the site of New Orleans.

There is an old chestnut of a joke about this aspect of English gardens. I suspect that the story goes back to the Middle Ages (at least), but the version I heard involved an exchange between the then premier of the Soviet Union, Nikita Khrushchev, and the head of one of the Oxford colleges. When Khrushchev asked how the gardeners had created such a perfect lawn, the don replied, with a sniff, "One plants it, waters it, fertilizes it, trims it, and then rolls it for five hundred years."

The point I am making is that managing ecosystems is not a new endeavor for human beings. It's one that has, at least at places like Merton in Oxford, been carried out for many centuries. And this thought is what led me to my suggestion, in the opening pages of this book, that the proper metaphor for the future relationship between human beings and nature is that of the garden and its gardener. Consider, if you will, the following points:

1) The gardener studies the garden and learns how it works.
2) The gardener does not "conquer" the garden, but uses his or her knowledge to get what he or she wants from it.
3) The gardener does not wantonly destroy parts of the garden, although he or she will pull weeds and other plants that aren't wanted.
4) The gardener plans the garden for a specific purpose, be it supplying vegetables, supplying flowers, or decorating a landscape.
5) The needs and wishes of the gardener shape the future of the garden, not vice versa.

In point of fact, modern humans have always behaved as gardeners toward some parts of nature. We saw how protoagriculturalists first observed the plants and animals around them, then arranged their lives to take advantage of them and, coincidentally, shaped the future of surrounding ecosystems. We discussed the way that Native Americans used fire and irrigation to shape the landscapes of the New World, and anyone who has known a farmer has probably noticed the sense of connectedness between the man and that small part of the global ecosystem that he manages. What I am proposing in this book, then, is that when human beings take the Second Step, it will not be a new departure for us, but simply a continuation of what we have always done, albeit with less knowledge and direct control than we will have in the future. I suspect that up to this point most of my environmentally conscious friends would not find much disagreement with my line of thought.

That changes, however, when we face the crucial issue, which is to ask how we will use our power to manage ecosystems in the future. The answer is not immediately obvious and does not concern science by itself. The opinion of scientists, in other words, carries no more weight in this discussion than the opinion of any other informed citizen.

There is certainly no shortage of candidates for principles that have been proposed to guide our interaction with nature. Some of these are too extreme to be taken seriously, and I won't look at them in any detail. We can't give up modern technological civilization (as some members of Earth First! might want), nor can we simply exploit natural resources without thinking about the consequences (a "business as usual" approach). Some more serious ideas making the rounds about what our management goal should be are:

1) We could manage to preserve endangered species.
2) We could manage to preserve biodiversity.
3) We could manage to preserve existing ecosystems.
4) We could manage to preserve the operation of natural selection in the biosphere.

The Endangered Species Principle

From the point of view of the public and the law, saving endangered species is the overriding concern of environmental management. Since the passage of the Endangered Species Act of 1973 (ESA), the EPA has been charged with designating species that are in danger of becoming extinct and formulating plans to preserve them.

The general idea of preserving species is based on the notion that each species on our planet carries a unique sequence of DNA. If the species goes extinct, that particular sequence of DNA is lost forever. Thus, good stewardship demands that we prevent extinctions from occurring, and that, of course, is the purpose of the ESA. But management based on this principle quickly gets us into hazardous territory. For one thing, there is the murky scientific issue associated with defining exactly what it is you are trying to preserve. If we take the uniqueness argument seriously, then it probably isn't just species we want to preserve, but some subspecies as well, and the task of delineating what is to be protected becomes difficult.

In biology, a species is defined to be an interbreeding population. Thus all humans are members of the same species *(Homo sapiens)* because we can all produce children with each other. Traditionally, the species has been the coin of the realm in discussions of population biology—after all, Darwin wrote about the origin of *species,* and one of our most important environmental laws is called the Endangered *Species* Act.

If you think back to our discussion of natural selection, however, you realize that the genetic makeup of populations is constantly shifting. In particular, new species develop gradually from old ones, with a slowly widening gap appearing between the genomes of the two populations. In the beginning of the process, there is only one species; in the end, there are two. What is there in between?

While this may seem like a brain teaser, in fact it is actually a familiar problem—I call it the problem of the continuum. Where along a continuum do you draw a line and say, "From here on, everything is X"? For example, there is little doubt that Bill Gates is a rich man and that a homeless person on the streets of New York is poor. But where on the income scale can you draw a line that divides rich from poor? No matter where you do it, you will always be able to argue that someone who makes one dollar more is really no different from someone who makes one dollar less.

Economists solve this problem by recognizing that there are grades of income—working class, middle class, upper middle class, and so on. Starting in the mid-1980s, conservation biologists realized that they, too, would have to begin making these kinds of distinctions.

The discussion actually started among people working at zoos. They realized that zoos could preserve only a small percentage of animals that are in danger of extinction in the wild (by some estimates, we might be able to save only about nine hundred kinds of birds, mammals, and reptiles, out of about forty-five thousand species of these animals in existence). In this situation, the question naturally arises: what gets saved? It's all very well to say we should save the tiger, but which tiger, exactly? The Siberian? The Sumatran? The Indian? To what extent are populations like these different enough to consider preserving separately, and to what extent are they in some sense interchangeable?

It was problems like this that led biologists to start talking about a concept called the Evolutionarily Significant Unit (ESU). The goal of this work is to find a way to designate the plants or animals on which scarce conservation resources should be spent. Since we can't save everything, in other words, we need to find ways of making policy decisions about what should be saved.

There are enormous consequences to seemingly esoteric debates about whether a particular population is an ESU. Depending on how the debate turns out, someone wanting to build a subdivision or an airport might or might not face legal hurdles and regulatory restrictions. A dam might or might not be built, depending on what kind of fish we decide we want to save, millions will or won't be spent on certain kinds of environmental research, and so on.

Unfortunately, the debate among biologists about how to define the ESU is far from settled. Basically, it comes down to an old problem between "splitters" and "lumpers." Splitters look at two things and see only the differences between them. The smallest detail, in their eyes, constitutes grounds for putting the two things into a separate categories. In their eyes, the fact that two groups of Pacific salmon return to different streams to spawn is enough to classify them as a separate ESU—a process that would split the salmon population into hundreds of different categories, each of which would have to be preserved.

On the other hand, lumpers look at two things and see the similarities. They see salmon from neighboring streams (which often accept each other as mates) as a single population.

So heated has this debate become that at a conservation biology conference a few years ago, scientists convened a special after-dinner session at a local pizza joint for a no-holds-barred discussion. The question remains unresolved, and the debate ended only when, after four hours, the exhausted staff threw them out.

In normal circumstances, this kind of scientific niggling wouldn't matter much outside the halls of academe. In today's litigious world, however, the obscure debate will become more and more important, because it will have enormous consequences for human activity. To take just one example, if you consider the Oregon and California spotted owl populations as separate ESU, you would have to take steps to preserve each separately. If, on the other hand, they are a single interbreeding population (as current data seem to indicate), the requirements for preservation could be relaxed considerably. The ability to carry out logging operations would be radically different in these two cases.

But assuming that we can resolve these issues, another set of questions arises when we consider how to justify the costs of species preservation to the general public. There are several general classes of arguments that have been used for this purpose—I call them the hidden treasure arguments.

They go like this: over the course of time, natural selection has produced a huge variety of genes in the plants and animals in the biosphere. Historically, almost all medicines useful to humans have come from tapping this resource. Medicines (which are themselves molecules) are derived from natural molecules that have evolved to meet other needs in other organisms. Advocates of managing the planet to preserve species then go on to argue that there are untapped resources in the biosphere—cures for cancer, cures for AIDS, and so on—that we will never be able to use if we let species go extinct.

As I mentioned in chapter 11, I have always felt that this is a particularly dangerous argument for environmentalists to make, because it puts species preservation at the mercy of technological advances. The advances in genomics, and particularly the development of designed drugs, may soon make the natural storehouse of molecules irrelevant to advances in medicine. We may, for example, have derived aspirin from the bark of the willow tree, but that is no longer a reason to preserve willows today—we already have what they could offer us. In the future, when advances in genetic sciences allow us to decouple medicine from the ecosystem, the hidden treasure argument will lose most of its force.

The Biodiversity Principle

A second guiding principle one often sees in environmental writings concerns not the preservation of single species, but the preservation of a wide variety of species. The term "biodiversity" refers to the number of different species that inhabit a particular place, and differs from the term "biomass," which is a measure of the total amount of living tissue in a place. The flocks of pigeons in a place like the Piazza San Marco in Venice, for example, represent a system with high biomass (because there are lots of pigeons) but low biodiversity (because there is only one species present).

The highest levels of biodiversity in the world occur in the tropical rain forests, and much of the discussion of biodiversity centers on those regions. Biologists have identified the regions under greatest stress—they are called hot spots—and they advocate concentrating our efforts on preserving the biodiversity there. In the long term, scholars like Harvard's E. O. Wilson look ahead to a time, perhaps in the twenty-second century, when human populations will begin to decline. "Our goal," he says, "is to carry as much biodiversity through [this population bottleneck] as possible."

As we saw in our look at the field of experimental ecology, in systems that operate according to the laws of natural selection, biodiversity acts as a kind of safety mechanism, providing a sort of insurance against bad times. An argument for preserving biodiversity can be based on this fact. If we want an ecosystem to survive (because it performs a useful service for humans, for example), then it behooves us to focus on maintaining the biodiversity of that system, so that it will weather the next drought or hard winter.

This argument makes sense up to a point, but only up to a point. For one thing, experimental ecologists have found that the advantages that accrue to many ecosystems as the number of species increases level off somewhere around ten species or so. Increasing biodiversity beyond that number doesn't seem to matter much.

Furthermore, this argument makes less sense in a managed ecosystem than it does in one that operates according to natural selection. If humans can intervene in times of ecological stress (as they would in the Merton College garden, for example), the need for a high biodiversity insurance policy becomes less. After all, every home owner knows that the way to shepherd the low diversity ecosystem we call a suburban lawn through a drought is not to increase the number of species of grass, but to water the grass that is there. In the same way, in a managed world, one would expect that choices between active management and maintaining biodiversity would be made on some sort of cost-benefit analysis, rather than according to a principle in which biodiversity is regarded as a good in and of itself.

The Ecosystem Preservation Principle

As scientific knowledge about the natural world has increased, we have come to realize that concentrating on the preservation of single species is a strategy that is profoundly out of tune with the way nature works. Virtually all of the extinction events in modern times are associated with the destruction of habitat—with the destruction of ecosystems themselves—rather than with events that affect a single species. Thus, the argument goes, the best way to preserve species is to preserve entire ecosystems. Insofar as such a diverse and cantankerous group of individuals as environmentalists can be said to have a mainstream opinion, this is probably what it is.

In some ways, the preservation of ecosystems is the hardest environmental policy to sell, because it requires the sequestration of large areas of land. I have heard people argue, for example, that to preserve the grizzly bear population in Yellowstone National Park, huge areas of Montana and Wyoming would have to be turned into a game preserve. (The argument is based on the fact that individual grizzlies can range over a tract of more than a hundred square miles.) Needless to say, the good citizens in areas bordering the park look on this kind of proposal with a decided lack of enthusiasm and suggest that there are other strategies (such as feeding the bears) that would stabilize the population.

Perhaps the most ingenious argument for ecosystem preservation involves a concept called ecosystem services. The idea here is that the world's ecosystems provide us with essential services—clean water, air, soil, and so on. If we put a monetary value on these services (by calculating what it would cost to provide those services artificially, for example), we can talk about the value of an undisturbed ecosystem and compare it to the expected value from whatever project would disturb it. For example, destroying a wetland area to build a shopping mall will generate a certain amount of wealth (in jobs, taxes, and so on). If, however, we have to build a water purification plant to take over the job that the wetlands used to do, it may turn out that the shopping center is a bad bargain.

The poster child for ecosystem services involves reservoirs in the Catskill Mountains in upstate New York—reservoirs that supply water to New York City. The myth surrounding these reservoirs is that development in the area caused the water quality to drop and that the city bought large tracts of land to avoid building a filtration plant. In fact, what drove the city's decision was not development (which was negligible in most of the reservoir area) or water quality (which hadn't changed), but new regulations from the Environmental Protection Agency. (One study even concluded that the biggest source of pollution in the area was the runaway population of beavers and whitetail deer.) Faced with a choice between spending close to one billion dollars to purchase land or spending six to seven billion dollars to build a filtration plant (with an annual operating budget of three hundred million dollars), the city—not too surprisingly—agreed to institute programs to maintain water quality. These programs included land purchase and subsidies to landowners for improved septic systems. As of February 2002, the city had purchased a little less than twenty thousand acres, at a cost of about sixty-four million dollars.

I have to say that after looking into the details of the story of the Catskills, I was rather less impressed with the argument for ecosystem services than I had been when I first heard the official story at a scientific meeting. It is, after all, one thing to respond to degradation of water quality that can be attributed to urban sprawl, quite another to respond to a bureaucratic diktat from Washington. Nevertheless, ecosystem services is an important concept, and one that will surely play a major role in future deliberations.

A better example of an argument for ecosystem services involves a series of events that occurred in southern Florida in 1969. There were plans under way to build a large airport on swampy land near Miami. Scientists looked at this area and noted that if the airport were built, salt water from the nearby ocean would infiltrate into the region's water table, contaminating water wells. In this case, the ecosystem (the swamp) was providing the service of maintaining enough water pressure to keep the salt water out. As it turned out, the cost of replacing the wells was greater than the expected benefit from the airport, and this was one of the reasons it was never built.

Another, somewhat more esoteric argument for ecosystem preservation goes under the name of the biophilia hypothesis. The basic idea here is that there is some sort of innate human attraction to complex natural ecosystems. We seem to prefer scenery that contains both water and a variety of plants and animals. As has been pointed out by E. O. Wilson, we like to sit with our backs to the cave wall, looking out at sources of food and water.

My encounter with the Black Hills buffalo described in the first chapter would certainly be an example of biophilia at work, as would the sense of "rightness" that many people experience in old-growth forests or tallgrass prairies. Nevertheless, I remain somewhat skeptical about how far this notion can be pushed as a basis for environmental policy. At a purely personal level, I have to admit that when I go into a tropical rain forest—the gold standard of the complex ecosystem—I don't get an emotional rush. In fact, my first reaction on encountering the heat, humidity, and discomfort there is, "Why would anyone what to preserve *this*?" I have observed this same reaction in enough people to suspect that the love of this particular complex ecosystem is an acquired taste, not something innate. (To be fair, I should add that my own favorite ecosystem—high desert—evokes the same negative reaction in many people. I suppose that's what makes the world go round.)

Another problem with the biophilia hypothesis from my viewpoint comes from observing what humans do when they build their own ecosystems. In the United States, there are hundreds, perhaps even thousands, of resort hotels where a lot of money is spent to provide a pleasant environment for guests. These places are usually characterized by artfully designed lakes, lawns manicured to within an inch of their lives, and lots of carefully tended (and carefully controlled) shrubbery and trees. They really don't have much resemblance to a wild ecosystem.

I can remember, for example, giving some lectures at such a hotel in Palm Springs, California. I had come in late at night, so I didn't really get a look at the place. When I walked out onto my balcony in the clear light of morning, however, I was amazed at what I saw. There, plunked down in the middle of the desert, was a lush oasis of green grass, ponds, and flowering plants. There were even mallard ducks swimming around! My first reaction was, "Yes, this is what God would have done if he'd only had the money."

The fact is that when human beings have the ability to construct any ecosystem they want, they build places like that hotel or the garden at my college at Oxford. This, it seems to me, argues against the existence of any inherent love of complex ecosystems, as well as against using biophilia as a basis for planetary management.

The Natural Selection Principle

Finally, a new idea starting to make the rounds among environmental theorists is somewhat more abstract than the principles that we have just discussed. Some ecologists, instead of focusing on individual species or even specific ecosystems, are suggesting that what's important is not so much *what* natural selection has produced as the *process* of natural selection itself. In this scheme, the idea is to preserve places on the planet where natural selection still operates over a broad area—the northeastern region of South America, for example.

I have to admit that I find this notion a bit too abstruse for a policy guideline. To save "evolution," we would probably have to deny human beings access to large swaths of the planet, something that would be very hard to justify in an increasingly crowded world. On the other hand, you can see this "lock it up and keep people out" philosophy operating in many places other than remote wilderness. I have, for example, seen it in the attitude of officials of the National Park Service in the western United States.

At a deeper level, however, I think this concept fails because it implies (implicitly, at least) that there are parts of the planet free of human influence today and that that freedom should be preserved. As many authors have pointed out, if we define "nature" as what happens when there are no humans around, then "nature" disappeared a long time ago.

A Better Way

All of the ideas described above represent real, well-articulated points of view in current environmental debates. As I have tried to indicate, I don't regard the justification for any of them as being particularly robust. None of them stand up to two simple questions—"Why?" and "So?" We should save endangered species? Why? Because extinction is forever. So? . . . You can see how this gets tangled up. In the end, every line of questioning forces someone proposing a principle to govern environmental management (including the author) to face the fact that ultimately all justifications for policy choices come down to deeply felt moral or ethical feelings—feelings that manifest themselves when you finally answer, "Just because!" In the end, there is no strong scientific or economic justification for these choices.

Indeed, my problem with most of the arguments presented in the standard environmental debate is that they focus almost exclusively on the nonhuman parts of nature. No matter which option we accept, the bottom line is that humans should manage the planet for the benefit of species, ecosystems, biodiversity, or the evolutionary process itself, for things other than themselves. Despite the occasional bow to human needs, as in ecosystem services or hidden treasure arguments, the real motivation driving most of the environmental movement, it seems to me, is the conviction that humans ought to curb their own needs and desires, to give their own needs less weight than those of nonhuman species.

It's not surprising, then, that the sorts of justifications advanced to support these viewpoints tend to be philosophical or even religious in nature. Humans have no moral right to cause the extinction of other species, we are told, because only humans have the ability to be aware of the consequences of their actions. We have a duty, we are told, to see that what natural selection has produced is passed on to future generations—to be good stewards of the planet. This list of moral principles can be expanded, of course, but the general import is clear. For whatever reason, the environment is to be managed for the benefit of nature.

I have become aware that these views are held by many educated people in the United States (including, I might add, my own adult children). They constitute a kind of quasireligious approach to the environment that many find satisfying. I would suggest, however, that they do not constitute the only kinds of moral or ethical standards that can be applied to this issue. I would argue that equal moral weight (as well as a great deal of practical expediency) can be attached to the notion that the environment should be managed for the benefit of human beings.

We can point to many historical examples in which ecosystems were managed for the benefit of human beings. Early agriculturalists learned to protect specific species because those species would supply food. Native Americans used fire to turn huge areas into what were essentially game parks. Modern farmers manage ecosystems to feed people worldwide. Surely feeding and caring for human beings is as lofty a moral goal as saving an endangered insect.

I introduced this way of thinking about management issues in what I called the benefit-to-humans principle in chapter 1. Here it is again.

*The global ecosystem should be managed for the benefit,
broadly conceived, of human beings.*

In the next chapter, I will talk about what a world managed by this principle might look like, but for the moment, I simply want to advance the proposition that the welfare of human beings is as good a moral or ethical principle as any of those listed above. Indeed, in my own personal moral calculus, it is a more worthy one.

To buttress this claim, let me refer back to the discussion of the role of DDT in saving human lives. You will recall that I asked you to play God by weighing the choice between using or not using DDT, with the knowledge that it would save ten million human lives, but cause serious problems for birds and other animals. Just to make the ethical distinctions sharper, let us suppose further that the choice was DDT or nothing—that chemists would be unable to develop an effective compound that was less environmentally damaging. This is what physicists call a thought experiment—an admittedly artificial situation designed to cast a particular issue into sharp logical focus. In this example, the choice is stark—do you save human beings or songbirds?

Aside from a few members of organizations such as Earth First!, I have never met anyone who would have trouble with this decision. The choice is obvious for most of us—ten million human lives clearly outweigh any damage to birds.

Now let's make things a little harder. Suppose we lower the number of human lives saved. How about one hundred thousand? One hundred? Ten? One? What if the only human value in the scale isn't actual survival, but quality of life? What if preserving the habitat for a particular species drives the price of lumber up enough so that a young couple can't afford to buy a house and has to raise their family in what they consider an inferior environment? What if saving a wetland means that someone has to leave his or her hometown because there are no jobs available? As you move down the scale from saving lives to providing for quality of life, the moral clarity of the original thought experiment starts to become murky, and the decisions become more complicated and less clear.

I suspect that when we start to manage the planet in earnest, there will be very few clear moral decisions. We will find that we have to make every decision on a case-by-case basis. As I tell my classes, "Save the Whales" is easy, but once the whales are saved, things get tough. We have not had to make these kinds of hard decisions up to now because we simply didn't have the ability to affect nature on a large enough scale to matter. With advances in science, this will change over the next few decades, and we have to start thinking now about how these decisions will be made.

Before moving on, let me address one other type of reaction I have often heard. When I described the notion of a managed planet to a friend, he responded by saying, "It makes me sad to think of a future where there are no free running rivers." It really didn't matter to him if he ever actually visited his river—he just wanted to know it was there.

I have since encountered this sort of response often enough to realize that it is a deeply felt conviction for many Americans. I suspect, for example, that a good deal of the opposition to drilling in the Arctic National Wildlife Refuge in Alaska comes from just this sentiment—we will never personally visit the part of the tundra involved, but we want to know it's there, untouched.

On the one hand, making people feel good can certainly be classified as a benefit to humans—in fact, it was in recognition of this kind of psychological benefit that I put the phrase "broadly conceived" into the principle. On the other hand, this kind of argument often takes on an unfortunate aspect of the elitism that mars some environmental writing. Removing a dam from a river may well return it to its free running state, but it may also cut off irrigation water or produce floods that could drive families off of farms that they have worked for generations. In this case, you have to ask which humans should get the benefit—the farmers who are directly affected or the urban environmentalist who gets to feel good. I asked my friend, for example, if he was willing to be the one to go to that family and say, "I'm sorry you're going to lose your farm, but it's important to me, sitting in my (Manhattan apartment/suburban home/retirement community in Arizona) to know that the river is running free." After some thought, he said that he wouldn't want to be the one to do it, although that really didn't change the way he felt.

I bring up the specter of elitism here because I often see signs of it when these sorts of arguments are advanced—not among scientists so much, but in the general debate. It takes many forms, from the "Not in My Back Yard" opposition of wealthy property owners to new developments to the sort of "I just want to know it's there" attitude of people who will, in all likelihood, never see the ecosystem that is the subject of debate. Current squabbles over the placement of windmills—devices that represent the alternative energy source most likely to compete with or replace fossil fuels—are an example of what I mean. Hidden behind the numbers and aesthetic arguments in these debates, I often think that I hear something like "We here in Cape Cod don't want to be forced to look at windmills on the horizon, but it's okay for those hicks in North Dakota." I hope that this will change, because this perception, whether justified or not, makes it hard to develop a working plan for our planet.

"Feel good" environmentalism is not at bottom a scientific issue, but one that grapples with the social and political trade-offs we will have to be able to make if we are to manage the planet. It is also precisely the sort of conflict that our political system is designed to handle. My hope (perhaps naive) is that we will learn how to conduct these sorts of debates with a modicum of rationality, and perhaps even civility.

15

The Managed Planet

We humans are a discontented lot. We have never been willing to take only what nature offered us—we have always wanted to improve on it. It's inconvenient to live in caves, so we learned to construct shelters, a skill that eventually led to modern skyscrapers. Hunting and gathering is a precarious existence, so we developed agriculture, which eventually led to the enormous human populations that now inhabit the Earth. The world is full of microbes and viruses that cause disease and death, so we developed medicine and found ways, in many cases, to eliminate these risks. Throughout the history of our race, in fact, we have always tried to improve on nature. To the fullest extent permitted by our technologies, we have modified our environment to make our lives better. The question before us, then, is not whether our next step marks some entirely new and unfamiliar departure, but what form this continuation of our ancient ways will take.

In point of fact, we have been managing ecosystems on the local level for millennia and at the regional level for decades. The states and Canadian provinces bordering on the Great Lakes have long-standing agreements about the use of water, for example. I have heard engineers in Chicago joke that they have to get permission from the Canadian prime minister before they can flush a toilet—an exaggeration, of course, but one that points to the interconnected set of regulations that govern the use of the water in the region. States in the Chesapeake Bay watershed are in the process of regulating runoffs into the bay, with an eye toward preventing further losses of productivity in that body of water. These sorts of rudimentary regional management regulations can already be found around the world—in the Danube Basin and in the rivers of the American Southwest, for example. What will change in the future is that these regional schemes will be linked together into a much larger, global effort, an effort that will lead us to the managed planet that I have talked about.

I think of this process as analogous to the building of the road system in the United States. In the beginning, roads were a local affair, the responsibility of each town and hamlet. Later, regional road systems, operated by state governments, became widespread. Finally, the interstate highway system was built, a national road system that both superceded and incorporated everything that had gone before. I suggest that we are at the brink of building that "interstate" system in the area of ecosystem management, as we make the transition from large-scale regional to global management.

Will this transition be easy? No—we still have a lot to learn about how to do it, a point to which I shall return below. Will the transition be free of mistakes? Of course not—as I have pointed out repeatedly, large-scale engineering projects always involve mistakes and corrections. But will it be something completely new? Not really—it will simply be an expansion of things our race has been doing for the last ten thousand years and as such will be just one more step for us.

As I look ahead to this future, I see a number of barriers. Some of these are technical, involving the acquisition of scientific knowledge we have talked about in the preceding chapters. Others are more psychological, involving firmly held ideas and preconceptions about the magnitude of the task before us—ideas that at the very least need to be examined in the light of new advances in science.

The most common reaction to the notion that we must manage the global ecosystem in the future is a vague sense of unease, a sense that we really shouldn't be tampering with nature in this way. Some of this hesitation is a natural human resistance to change, but some of it derives from a belief that nature, left to itself, really won't do anything bad to us. If we just let nature work its magic, the argument goes, things will be pretty much okay.

There are several ways to approach this argument. One is simply to point out that, left to itself, nature can (and will) produce a lot of bad outcomes. It is perfectly natural for rivers to flood, for example, driving people from their homes and causing huge property damage. It is profoundly unnatural to control floods by building dams and controlling river flow, but the result is a much better outcome for human beings living near rivers. Nature, left to itself, produces hurricanes, blizzards, tornadoes, earthquakes, and volcanic eruptions, none of which can be controlled at present (although their damage can often be minimized by the advanced warnings supplied by modern technology). Nature, left to itself, surrounds us with bacteria and viruses that cause suffering and disease, and the effects of these agents can only sometimes be countered or managed by modern medicine. Leaving things to nature does not guarantee a pleasant or even an acceptable outcome for human beings. Had we followed the "Let Nature Do It" line in the past, chances are that very few people reading this book would be alive today. (For that matter, the "unnaturalness" of both printing and reading might even have made reading this book impossible.)

Of course, human interventions (the building of dams, for example) have had a bad effect on various aspects of the natural world—on the survival of an endangered species or the displacement of a habitat by farmland. In cases where human benefit is obtained at a cost to the environment (as it usually is), we have to ask what is more important—the benefit to humans or the cost to another species. As I argued in chapter 14, human welfare is at least as good a moral justification for action as any of the (somewhat doubtful) principles advanced for environmental protection.

But there are those who argue that humans simply lack the ability to manage the environment, mainly because of a deep distrust of human political institutions. Most often, however, this distrust is based on some version of our old friend, the law of unintended consequences. No matter how careful we are, the argument goes, we won't be able to think of everything, and sooner or later something disastrous will happen.

The key point to recognize here is that deciding to let nature do it is, in and of itself, a human decision. It's no good pretending that every human decision is subject to this law *except* the decision to allow nature to take its course. As I argued in chapter 13, the decision to build a dam may well have unintended consequences, but so may the decision not to build a dam. Letting nature do it may make some people feel good by seeming to relieve them of the responsibility of making a tough decision, but it doesn't change the fact that a decision must be made. It may well be that in some cases future managers of the planet will decide to step aside and let nature supply us with a particular service (clean water, for example), but they should do so only after a detailed analysis of the situation shows that this is the best course to take, not because of an a priori assumption about the benevolence of nature.

Actually, I think that unease about managing the planet (a feeling that, incidentally, I share) is similar to the feeling many of us have experienced when we suddenly realize that we are going to have to make decisions about the care of our parents. Not only are they not going to take care of us anymore, but now we have to start taking care of them. In the same way, "nature" can't be taken for granted anymore and has become our responsibility. Neither of these is an easy transition for any of us to make.

Trantor versus Utopia

There is a story that circulates in the environmental literature that I have heard so often, and in so many guises, that I am beginning to suspect that it is a modern urban legend. It goes like this: a child, usually a ten- to eleven-year-old girl, is told about plans to build pollution-free cars or electrical generating plants. She expresses dismay and, when asked why, replies, "But that means that they can just pave everything!" (I should say that some of my suspicion about this story arises from the fact that two separate colleagues, on opposite coasts, assured me that it was their daughter who was the child involved.)

Will the advent of the managed planet really be a boon to paving companies? Would a managed future necessarily be one in which more areas of nonurban spaces were paved over to make way for cities and suburbs? That is certainly one possible outcome, but, to my mind, an unlikely one. When I think about what a managed planet would look like, I think of a range of possibilities on a continuum from total urbanization to a total return to "nature."

The first option was imagined in stunning detail by Isaac Asimov in his science-fiction classic *Foundation*. Set in a far distant future, when human beings had colonized the entire galaxy, the story introduces the mythical planet Trantor, home of the galactic government. In Asimov's words, "All the land surface of Trantor . . . was a single city. The population . . . was well in excess of forty billion." In this imagined future, then, a world is indeed "paved" in ways that present-day earthlings can't even imagine.

A contrasting view has been put forward by Harvard ecologist E. O. Wilson, who imagines a future in which a population of a few hundred million humans live in a Utopia in which the forces of natural selection produce the kind of biodiversity that would exist without humans, although Wilson's select few would presumably have enough of the benefits of modern technology to live comfortably.

I suppose it will not come as a surprise to any reader who has come this far with me to learn that I don't think either of these outcomes is very likely. Even leaving aside the issue of where food would be produced on a paved world (Asimov had it shipped into Trantor from other planets), simple laws of economics, it seems to me, would militate against either extreme. For example, if future governments decided to follow a "pave everything" policy, remaining open space would become more and more scarce and, therefore, more and more valuable. The added benefit to be derived from the next square mile of paving would drop relative to the value of the open space itself, a fact that would effectively stop the paving at some point. In parts of the United States today, particularly in the West, this point may already have been reached.

Consequently, the managed planet is probably not going to look too different from what we find in developed countries today. There will be areas of intense urbanization, of course, but I suspect that then, as now, most of the land area of the planet will be open. I have found, for example, that the best way to convince my environmentally conscious friends that the country is not being paved over is to talk them into taking a long drive, particularly through places like Wyoming or eastern Montana. To easterners, driving for hours through open rangeland is a powerful and illuminating experience, illustrating the fact that North America is still almost all open space. This, I think, is part of what lies behind the current efforts of chambers of commerce in Montana to market the state as the "Last, Best Place."

But the real point about the second step is that what the world will look like in the future is a decision that we, collectively, have it in our power to make. We can, if we want to, choose any spot on the continuum from Trantor to Utopia and say, "Here—this is the world we want," and then go ahead and create that world. My guess is that that world will include both cities and wilderness areas, places full of people and places where people can get away from each other.

Managing Well

Another reaction I often get when I discuss the notion of a planet in which humans take control of nature is a profound skepticism that humans actually have the ability to manage the planet. "Look at how politicians have screwed things up already," the argument goes. "How can you believe that we won't just screw this up as well?"

Don't get me wrong—I enjoy trashing politicians as much as the next guy—but I think this attitude just won't fly. In fact, once we started paying attention to the environment, we've done a reasonably good job of cleaning it up (although the job is nowhere near completed). There was a time, for example, when the Cuyahoga River in Cleveland was so loaded with industrial waste that it actually caught fire—an event that a local microbrewery has commemorated with its Burning River Ale. Today, game fish are returning to many urban rivers in the United States as water quality improves and, as I pointed out in the first chapter, the air quality in our cities has improved as well. Despite the constant drumbeat of gloom from the environmental press, there are many success stories in environmental management.

This fact was brought home to me a few days after my encounter with the Black Hills buffalo that started the train of thought that led to this book. I was sitting on a motel lawn on the shores of Lake Superior, near the Apostle Islands in Wisconsin. I was reading a local bird guide (published three years earlier) that described the bald eagle as a "rare" sighting, when a huge eagle flew up and perched on a dead tree nearby. Since that time, I have seen these magnificent birds in many places, including a young pair near my favorite roller-blading track near Reagan National Airport on the Potomac River in downtown Washington, D.C. Since the bald eagle was one of the species most affected by the use of DDT, their return can be seen as another success story for intelligent management.

Having said this, I have to add that there are, of course, examples of poor management. Our current problems managing forest fires, for example, is largely a result of the decision not to clear forests in the past while at the same time suppressing fires. My impression, however, is that these days bad decisions are as likely to be made by ideologically driven environmentalists as by rapacious industrialists. Let me give you just one small example from my recent trip to national parks and wilderness areas in the West.

My favorite hiking trail in the world winds up the flanks of the Beartooth Mountains in Montana, going through a wilderness area on its way to a plateau at ten thousand feet. At one point, the trail goes through a high mountain valley that contains lakes and large, swampy expanses of low willow. It's easy to get lost in that valley, so over the years, hikers have built up stone cairns to mark the trail—nothing high tech, just some piles of stone to keep hikers from getting lost in the swamp.

On my most recent hike, I found that the cairns had been removed. Someone in the Forest Service had decided that the "unnatural" piles of stone didn't belong in a wilderness area and had torn them down. How long would it be, I wondered, before a newcomer to the trail wandered off into the swamp, became injured, and died? What a price to pay for ideological purity! So as we move toward a managed planet, we must be careful to avoid excesses from both sides of the political spectrum.

When it comes to predicting exactly what sorts of institutions will be needed to run the management process, however, I'm afraid that we have moved outside of my own particular area of expertise. From the nature of the problems we face, however, it is clear that some of these institutions will have to be transnational, perhaps evolving out of current organizations like those monitoring the Montreal and Stockholm accords, perhaps created ab initio, like the U.S. Department of Homeland Security. Like everything else in this messy world, these institutions will evolve, make mistakes, correct them, and go on. They will, in short, be human, but, in the end, they'll get the job done.

The Future Managed Earth

But the big question still remains: what will a managed planet look like, ecosystem by ecosystem? A project now under way in Florida offers a preview of how human management of nature might look in practice.

The Florida Everglades stretch over the southern tip of the Florida peninsula. They constitute a strange ecosystem. From a distance, they look like prairies punctuated with groves of trees but on closer examination are revealed to be almost entirely covered with water. The Indian name—River of Grass—sums up the nature of the Everglades. Created about five thousand years ago by overflow from Lake Okeechobee, the Everglades act as a kind of giant sponge located at the southwestern end of Florida. They are a kind of filter, cleaning water that flows, ultimately, into the Gulf of Mexico. Because of its geology, it also supports a unique ecosystem, a fact that was recognized by the formation of Everglades National Park in 1947.

Since the 1920s, important changes have occurred in Florida that have altered the Everglades. Farmers, particularly citrus and sugarcane growers (a quarter of the nation's sugar is grown in this region), have taken over large tracts of land, diverting water that used to flow into the Everglades. A system of dams and levees was begun at that time and was further expanded in the postwar period in response to a series of disastrous floods, so today there are already seventeen hundred miles of canals to move water around the region. Developers drained low, swampy land to make room for cities and subdivisions, as Florida was transformed from a sleepy rural backwater to a state with one of the highest populations in the country. Over the decades, the water that once supplied the River of Grass began to dry up as other uses were found for it. Water in large quantities is generally measured in a unit called the acre-foot—about enough water to cover a football field to the depth of one foot. In 1900, about 1.6 million acre-feet of water flowed through the Everglades each year. Today, that number is about 800,000 acre-feet—half of what it was at the start of the twentieth century.

By the 1980s, the reduced supply of water and the conversion of swamplands to developments and agriculture had progressed to the point that the effects on the Everglades were apparent, and the alarm was sounded. Something had to be done, but what? Was it more important to save the unique Everglades ecosystem? To keep the Florida development industry humming? To keep the sugarcane fields productive? This is, in fact, a classic example of a modern environmental dilemma. There are many constituencies, each with valid arguments to back up its claim, each with a compelling human benefit to be served by using science to manage the environment. All of them can't be satisfied, so some sort of compromise has to be worked out.

In 2000, the Comprehensive Everglades Restoration Plan was launched by a combination of federal and state officials. It was an unusual political process, with lobbyists for the Audubon Society and the sugar growers lobbying congressional offices together. In essence, the water flowing through Florida was divided up—so much for developers, so much for farmers, so much to keep the Everglades alive. How successful the eight-billion-dollar project will be in balancing all of these needs remains to be seen, of course, particularly because there are a lot of new and untried water-storage technologies involved in the plan. From the fact that all three parties are complaining about the agreement, however, I suspect that the negotiators have come about as close to finding an ideal compromise as was possible. Even though it's still in its infancy, the plan is serving as a model for large-scale projects from Brazil to Africa.

So the Florida Everglades—the epitome of the wild swamp, of untamed nature—will survive. It will do so, however, as part of a huge engineering project that will, in essence, turn the Florida peninsula into one gigantic plumbing system. Will the Everglades remain "natural"? By the definition I introduced in chapter 2, where "naturalness" depends on whether or not natural selection operates, they will be "natural." They won't, however, be able to function independently of human intervention, and in that sense they will be managed.

In fact, the Everglades Restoration Plan is a kind of blueprint for what a managed planet will look like. You will be able to take your kids to see the alligators and the herons, but somewhere in the background, there will be an engineer opening and closing valves, keeping the ecosystem running the way we want. The Everglades, then, represent a near-term vision of management, one that we will see playing out over the next decade. There are many longer-term visions as well, and I will talk about one of them, just to give you a sense of what they look like.

Shaping Ecosystems to Meet Human Needs

In chapter 9, I discussed the enormously complex problem of global warming and touched briefly on some of the policy options available for dealing with it. There are, of course, many options available beyond the rather conservative choices there. One—an option that has a faint whiff of science fiction to it—goes by the name of geoengineering. Just as the developers of agriculture decided not to be content with what nature offers in the way of food, geoengineers do not wish to be content with what nature offers in the way of a planet. Instead, they want to shape the planet to fit human needs.

In its extreme form, this impulse becomes the futuristic notion of terraforming—of turning a planet like Mars into a place with an atmosphere and oceans, like the Earth. In its less ambitious form, it becomes a search for ways to modify the Earth itself.

In 1987, the late oceanographer John Martin of Moss Landing Marine Laboratories in California noted an important fact about the Earth's oceans. Over large areas, particularly in the southern Pacific, the growth of ocean plankton is limited primarily by the availability of iron. Since most of the photosynthesis on the planet takes place in the oceans (not, as commonly believed, in tropical rain forests), the scarcity of iron is what limits the ability of plankton in these regions to pull carbon dioxide out of the air and incorporate it into their skeletons. (While talking about plankton may seem exotic, this process of pulling carbon dioxide out of the atmosphere is actually quite familiar. A tree does the same thing—it takes the carbon dioxide from the air and incorporates it into wood fibers and leaves.)

Martin's idea was simple: if you add iron to the ocean, you are doing the equivalent of fertilizing a cornfield. The iron would be taken up by plankton, which would multiply wildly, pulling carbon dioxide from the air and reducing the greenhouse effect. "Give me a tanker full of iron filings," Martin boasted, "and I'll give you a new Ice Age!"

The first ocean fertilization experiments in 1993 didn't work very well. Basically, the swiftly moving surface currents pushed the iron too far down beneath the surface to increase the number of plankton as much as lab experiments had predicted. As an aside, I should mention that it was watching environmentalists gloating over this result at scientific meetings that made me develop such a negative attitude toward some of their thinking. "See," they seemed to be saying smugly, "you aren't going to be able to fix the greenhouse effect. You're going to have to stop driving SUVs and live just the way I think you should, whether you want to or not." It was almost as if the prospect of finding an easy way out of the greenhouse problem was more than they could bear.

Fortunately (or unfortunately, depending on your point of view), subsequent experiments were more successful. Led by Ken Coale of Moss Landing, a team including thirty-eight scientists from thirteen universities and research centers figured out how to inject the iron in smaller doses, using satellite data to follow a single patch of water as it moved across the ocean. Their result was little short of astonishing. In a few days, the sea turned green around the ship, and the nets used to bring samples aboard gummed up with green scum. "It was like sailing into a duck pond," Coale said at the time.

All in all, this experiment wound up pulling about a hundred tons of carbon out of the air. (For reference, your car puts about a ton of carbon into the air in a year.) Following this triumph, funding for the project was cut off—a move that administrators at the National Science Foundation say was not made for political reasons—but has recently been resumed. At the moment, the main scientific hurdle in this branch of geoengineering has to do with the question of whether or not the carbon that was taken from the atmosphere will stay in the ocean or whether it will simply be returned to the atmosphere. The hope was that when the plankton die, their skeletons would sink to the bottom where they would eventually be incorporated into limestone. It appears, however, that other biological processes are at work, and at least some of the carbon stays near the surface (after being eaten by fish, for example), which means that it might have an easy path back to the atmosphere. This kind of issue, together with other issues relating to the effect of increased plankton production on the ocean ecosystem, will have to be resolved if ocean fertilization is to have a place in humanity's future management of the planet.

I expect, however, that as we get more familiar with the way the Earth's ecosystem and climate work, more kinds of geoengineering projects, both large and small scale, will be proposed and developed. Some of them will undoubtedly fail to pan out, but others will, and in the end, we will have a constellation of effective management techniques. Perhaps the most intriguing comment about this aspect of our future came from University of Chicago climate scientist Ray Pierrehumbert, whose work on global warming is discussed in chapter 9. Thinking about the prospect of controlling the climate, he said, "I can see a debate in the Earth Parliament ten thousand years from now about whether or not we will let the next Ice Age happen."

That would be some debate to watch!

Engineering Species to Meet Human Needs

But the real joker in the pack of future planetary management is genetic engineering. Except for the debate about the safety of the food supply discussed in chapter 11, my sense is that very few people involved in the environmental debate have given serious thought to the new science of genetics. The ruling assumption seems to be that however much human beings change the biosphere, we will always have to work with the same basic genetic abilities that natural selection has produced.

For example, one aspect of the greenhouse problem that has received a lot of attention has to do with the fact that the limiting resource for plant growth in most parts of the world is the availability of carbon dioxide (which is, after all, a relatively rare gas in the atmosphere). Increasing the carbon dioxide, it has been argued, will stimulate plant growth, which will, in turn, pull carbon dioxide out of the atmosphere.

While it is true that this sort of stimulated growth occurs in the laboratory, there is a question about whether ecosystems (forests, for example) can be counted on to do so in real life. At the moment, the data are unclear, but it seems that in at least some ecosystems plants exposed to increased carbon dioxide concentrations experience a growth spurt, but quickly adjust to the new conditions and go back to normal levels of carbon sequestration. How these studies will play out and how much carbon plants will be able to absorb are still open questions, but what intrigues me is that everyone seems to be operating on the assumption that the trees we have around today are the only kinds of trees that will ever exist.

It is possible, however, to imagine a different kind of future. Suppose that when we start to understand the genomes of plants, we find the genes that govern the uptake of carbon dioxide. In particular, suppose that we find the gene or genes that turn down the carbon uptake when atmospheric concentrations go up. It's not too hard to imagine that we could engineer plants that keep taking in more carbon, regardless of what's happening in the atmosphere. These plants would suffer a selective disadvantage compared to plants in the wild, since they would by definition have a less efficient carbon metabolism. They would have to be planted (and perhaps tended) by human beings, but they could perform a useful function in the area of climate control if we decide that we want to lower the greenhouse effect.

It doesn't take much imagination to jump from there to other kinds of "second generation" genetic modification that would have a profound effect on environmental management. You can imagine plants designed to pull nitrogen out of the environment, for example, and scientists are already trying to develop microbes to clean up oil spills. This is all pretty futuristic, of course—we don't know enough about either genomics or ecology right now to do it. Nevertheless, it's a possibility that we can't afford to ignore. I call it the "joker in the pack" because (a) we don't really know what is possible, and (b) we just haven't thought much about it. On the other hand, if we take the metaphor of human beings as gardeners seriously, we should recognize that humans have been planting and caring for modified plants for thousands of years. It wouldn't be too surprising, then, if we kept on doing so in the future with the next of our new genetic technology.

So ready or not, like it or not, new advances in science will make us the gardeners of the planet. Nature, in the sense of wild places outside the realm of human care, will cease to exist. Our ancestors took the first step along this road thousands of years ago, when they began to remove themselves from nature. It is now our turn to take the next step, to return to a nature that will be more human, more managed, than it has ever been. It is our responsibility to decide what kind of world we want to live on, because we now have the power—and the responsibility—to bring that world into existence.

To those who believe that humans just don't have the ability to carry out this task, I would point out one thing. To turn away from the opportunity in front of us involves, in effect, making a bet against human ingenuity and intelligence.

Historically speaking, my friends, that is just about the worst bet anyone can make.

INDEX

wilderness. *See also* nature
   cairns and, 227
   early humans and, 81–89
   myth of pristine, 79–81, 90–91
   poor management and, 227
Wilson, E. O., 212, 215, 225
wind power, 47, 145, 220
wind tunnel, 191
Wofsy, Steven, 140

Wolfe, Tom, 195
Wyoming, 225

yellow fever, 49
Yellowstone National Park,
   213
Yucatán Peninsula, 85

zooplankton, 107

ABOUT THE AUTHOR

James Trefil is the Clarence J. Robinson Professor of Physics at George Mason University. He has been a contributor to *Smithsonian* and *Astronomy* magazines and a frequent commentator for National Public Radio. He is the author or editor of over thirty books, including *Are We Unique?* and the bestselling *Dictionary of Cultural Literacy.* A fellow of the World Economic Forum and a Phi Beta Kappa visiting scholar, he lives in northern Virginia.